PROTOPLASMATOLOGIA

HANDBUCH DER PROTOPLASMAFORSCHUNG

HERAUSGEGEBEN VON

L. V. HEILBRUNN UND **F. WEBER**

PHILADELPHIA GRAZ

BAND IV

VIRUS

3

THE MULTIPLICATION OF VIRUSES

4a

VIRUS INCLUSIONS IN PLANT CELLS

4b

VIRUS INCLUSIONS IN INSECT CELLS

5

ANTIBIOTIKA ERZEUGENDE VIRUS-ÄHNLICHE FAKTOREN IN BAKTERIEN

SPRINGER-VERLAG WIEN GMBH

1958

THE MULTIPLICATION OF VIRUSES

BY

S. E. LURIA
URBANA/ILLINOIS

VIRUS INCLUSIONS IN PLANT CELLS

BY

KENNETH M. SMITH
CAMBRIDGE

WITH 5 PLATES

VIRUS INCLUSIONS IN INSECT CELLS

BY

KENNETH M. SMITH
CAMBRIDGE

WITH 16 FIGURES

ANTIBIOTIKA ERZEUGENDE VIRUS-ÄHNLICHE FAKTOREN IN BAKTERIEN

VON

PIERRE FREDERICQ
LÜTTICH

SPRINGER-VERLAG WIEN GMBH
1958

ISBN 978-3-211-80488-9 ISBN 978-3-7091-5461-8 (eBook)
DOI 10.1007/978-3-7091-5461-8

The Multiplication of Viruses[1]

By

S. E. Luria

Department of Bacteriology, University of Illinois, Urbana, Illinois

Dedicated to Professor Giuseppe Levi on his 85th birthday, as a token of respect and affection

Contents

	Page
Problems of Virus Multiplication	2
Multiplication of Bacterial Viruses (Bacteriophages)	4
Early phases of phage-cell interaction	4
The vegetative reproduction cycle	6
Lysogeny	17
Multiplication of Viruses in Cells of Vertebrate Hosts	21
Properties of the viruses	21
General course of animal virus growth	25
Stages in virus-cell interaction	26
Fate of the host cells in virus multiplication	34
Multiple and mixed infection	35
Persistent virus-host association in vertebrate cells	37
Multiplication of Viruses in Insects	40
Insect-pathogenic viruses	40
Viruses of vertebrates and of plants in their insect vectors	41
The CO_2 sensitivity agent of *Drosophila*	42
Multiplication of Plant Viruses	43
The nature of the infectious material and the initiation of infection	44
Overall course of virus multiplication	46
Persistent infections with plant viruses	51
The relation of the virus to the cell	51
References	54

[1] Aided by grants from the American Cancer Society, Inc. and the University Research Board, University of Illinois.

List of abbreviations

AMPS = d-Amino-2-p-methoxyphenyl-methane sulfonic acid
APC = Adeno-pharyngo-conjunctival (viruses)
CAM = Chorioallantoic membrane
DNA = Deoxyribonucleic acid
DNase = Deoxyribonuclease
FPV = Fowl plague virus
HA = Hemagglutinin
HMC = 5-Hydroxymethylcytosine
MNI = Mumps, Newcastle, influenza (group of viruses)
NDV = Newcastle disease virus
pfu = Plaque forming unit
RBC = Red blood cells
RDE = Receptor destroying enzyme
RNA = Ribonucleic acid
RNase = Ribonuclease
TMV = Tobacco mosaic virus
UV = Ultraviolet
WEE = Western equine encephalomyelitis

Problems of Virus Multiplication

Viruses are recognized operationally as infectious agents by their ability to produce recognizable alterations in living cells and tissues. The field of virus biology, and especially of virus multiplication, reflects the methodological consequences of the operational definition of viruses as infectious agents. Thus, evidence of virus multiplication requires evidence of the production of infectious virus in increased amounts. Even when we study the reproduction of viral materials in a more-or-less persistent noninfectious form (such as the "prophage" form of bacteriophage) we depend, for proof of the presence and multiplication of the viral material, on the ultimate production of some infectious virus. Only occasionally can we infer complete or partial phenomena of viral multiplication by indirect observations, such as multiplication of elements resembling known virus particles in morphological or physico-chemical properties.

Generally, we study virus multiplication by combinations of methods. The essential method is *titration,* that is, determination of the amount of infectious virus measured in multiples of a minimum (or convenient) amount called an *infectious unit.* The infectious unit may be the amount needed to produce one lesion; or to produce 50% positive responses; or to give a positive response within a certain time limit, under standardized test conditions. The infectious unit may correspond to one material virus particle, as for many phages (Luria et al. 1950), or to several millions of virus particles. It may reflect the initial action of a single particle or the chance combination of the presence of an adequate number of virus particles in a sensitive region of the host (see Beard 1956). Once a reproducible "infectious unit" has been defined, all titers can be compared on the basis of dilutions and titrations.

Viral multiplication is strictly intracellular. Hence, we must consider as its essential process the production of viral materials within virus-infected or virus-carrying cells. As already stated, this process is generally observable only insofar as it culminates in the production of infectious virus. It is most easily studied when it does so within a relatively brief span, namely, within the very cell that has been infected, without intervening cell multiplication. Even then, however, viral multiplication is a process radically different from the multiplication of cellular organisms, whether free living ones or intracellular parasites. We shall show that viruses multiply not as organisms but as subcellular entities, more on a level with organized, non-independent cell constituents. The dependence on the host is probably not a purely nutritional one, as for other parasites, but an integrative one. We may speak of viral multiplication as viral biosynthesis, in the same sense as we speak of the biosynthesis of macromolecular cellular constituents and organelles. This viewpoint need not rob viruses of their "independence," but interprets it as a genetic and evolutionary independence, limited by the demands and the implications of the integration process within the host cell.

Upon entering a susceptible cell, the viral material undergoes transformations (whose cytochemical basis is well understood only for bacterial viruses, but whose occurrence may be more general) which generally lead to a disappearance or "eclipse" of infectious virus detectable by infectivity tests on extracted cell contents. Most evidence indicates that the actual synthesis of virus material, and the multiplication of the viral elements as genetic units, takes place in the eclipse period. The virus multiplies in a noninfectious, *vegetative* form, and infectious virus particles are formed by a terminal process of *maturation,* in which viral materials become incorporated into individual infectious particles. Hence, the problem of viral multiplication is translated into a series of problems: transformation from infectious into vegetative form, vegetative multiplication, and maturation. The tools of biochemistry, immunochemistry and genetic analysis are required for the study of synthesis of viral constituents during vegetative reproduction.

Most intriguing situations are encountered when virus material, having once entered a cell, remains in noninfectious form while the cell proceeds to reproduce. The persistence and continuous multiplication of virus in noninfectious, *provirus* form within these cells is revealed by its occasional transition to the vegetative and mature, infectious forms in some of the descendant cells. This situation has been clarified only with phage, where the provirus or *prophage* can even be located within the linear genetic structure (chromosome?) of the bacterium. The possible role of noninfectious proviruses in other persistent virus-host cell associations in plants, vertebrates or insects, has not yet been clarified.

The intracellular mode of reproduction of viruses reduces most of the pathology of virus disease to the expression and amplification of changes in the structure and functions of virus-infected cells. Cellular pathology, besides revealing the presence or absence of viruses, can serve the study of viral multiplication. Viral materials may be traced morphologically (especially by electron microscopy) and cytochemically as abnormal or deviant components of cells or of cellular fractions, such as mitochondria, microsomes and isolated nuclei.

Once virus maturation has occurred, infectious virus may spread to other cells if means for release of infectious virus exist. The cytochemical basis of such release processes, which sometimes entail complete cell disintegration, sometimes may be compatible with continued cell life, remains one of the least understood phases of virus biology. In exceptional cases, possibly including some of the proliferative responses to viruses, viral multiplication and transmission might conceivably be purely intracellular, within the progeny of an initially infected cell, without any extracellular release or spread to other cells.

External control of viral multiplication may be exerted either by destroying the infectivity of mature virus particles (in their extracellular or intracellular environment) or by interfering with viral reproduction. Selective interference with reproduction—for example, for purposes of chemotherapy—has been harder to achieve for viruses than for other parasites, because of their lack of independent metabolism and of the similarity between viruses and basic host-cell constituents, as far as synthetic processes are concerned. Processes peculiar to viruses, such as those involved in virus maturation, may offer a more sensitive target for selective attack.

In this chapter we shall discuss first the processes of reproduction of bacterial viruses, then those of viruses attacking vertebrate hosts, insects, and plants. Our present knowledge of the mechanisms of multiplication of various groups of viruses decreases in approximately the same order. Finally, we shall compare the multiplication of viruses with that of other biological entities and attempt to define the position of viruses in the biological world. It is not the scope of this chapter to provide a critical review of these many fields for the specialist. Rather, we shall aim at describing to biologists in fields other than virology the present-day state of our knowledge of viruses as biological units of reproduction.

Multiplication of Bacterial Viruses (Bacteriophages)

The interaction between bacteriophage and host bacterium has been studied on a number of systems. Most information relevant to virus multiplication concerns certain groups of coli-dysentery phages, mainly the T phages, especially the T-even numbered viruses (see LURIA 1953 b, HERSHEY 1956); phage λ (JACOB 1954); and a few phages active on *Salmonella, B. megaterium* or *Pseudomonas.* We shall attempt to give a composite picture, assuming generality of mechanisms when it is plausible and not contradicted by known facts.

Early phases of phage-cell interaction

Attachment and penetration. Particles of many and possibly all phages consist of protein and DNA. Most phages have particles differentiated in a head and a tail; the head is often polyhedral or spherical in shape. The tail tip, when present, is the organ of attachment to specific receptor sites, which are present in large numbers on the cell wall of the host, so that many particles can attach to one bacterial cell (see TOLMACH 1957). Attachment is specific; it can be prevented by genetic changes or by blocking of the specific sites of either virus or cell by means of antibody. Attachment is also sensitive to ionic environment and, in some instances, to the presence of cofactors (such as L-tryptophan) required to activate the attachment sites of the virus (see STENT and WOLLMAN 1950). These requirements are specific for each system.

Reversible attachment, demonstrable under special conditions, is rapidly followed by an irreversible attachment. The structure of the phage particle (taking coli phage T 2 as a model) is deeply altered; the external part of the tail contracts, while an inner rod seems to project through the cell wall (KELLENBERGER and ARBER 1955). The phage DNA leaves the phage head and penetrates the protoplast (HERSHEY and CHASE 1952) leaving outside an exoskeleton containing over 90% of the virus protein. The empty protein bag can be removed from the cell surface without altering the course of infection. The process of injection is very rapid in some instances; in others, it is slower and requires Ca^{++} ions (LURIA and STEINER 1954).

The mechanics of this remarkable process of penetration appears to be as follows (KOZLOFF et al. 1957). Following attachment, an enzyme present in the phage tail is exposed, apparently by the action of a zinc-containing enzyme of the cell receptor, which removes the tail tip. The phage enzyme digests some parts of the cell wall. Finally, injection of the DNA may be triggered by entry of amino acids into the phage.

The selective penetration of viral DNA with little or no protein into the cell is a central fact in phage biology. It explains the eclipse of infectivity. It relegates most of the phage protein to the role of a carrier or "syringe" for the essential viral constituents. It makes the viral DNA the leading candidate for the role of genetic controller of reproduction, a view which agrees with evidence from transformation phenomena in bacteria (AVERY et al. 1944) and with the biochemical and genetic events in phage multiplication (see below).

Initiation of viral infection. Following the injection of DNA, there occurs in the bacterial cell a series of processes that may lead to any one of a number of alternatives.

1. The phage may undergo *vegetative multiplication,* leading to virus maturation and cell lysis (*productive response*) (DOERMANN 1953).

2. The cell may survive, multiply, and give rise to *lysogenic* progeny, that is, to cells that are carriers of a *prophage,* which has all the genetic potentialities of the infecting phage (*reductive response*) (LWOFF 1953).

3. The cell may survive and may retain and transmit to its progeny the infecting phage material with little or no multiplication (preprophage state, BERTANI 1953). This *immune response* is observed in superinfection of lysogenic bacteria with some phages related to their prophage; the preprophage can often interact with the prophage.

4. The infection may be "abortive" and the viral material may fail to multiply and persist in any recognizable form, either because of being itself damaged (for example, by irradiation) or because of being frustrated by a genetically or metabolically incompatible environment (ADAMS 1954). In some systems, this abortive infection by failure to initiate

reproduction leads regularly to death of the host cell (as with the T-even coliphages); in other systems the cell survives and apparently eliminates the aborted phage.

The decision as to which response will occur depends on host and phage heredity (including presence of prophages) and on environmental factors. Some phages, like the T-even and T 5, give only responses 1 or 4 (vegetative reproduction or abortive infection). These are supervirulent or *intemperate* phages. Other phages, which probably constitute the great majority, are *temperate* phages, which can elicit any one type of response depending on conditions. Typically, a temperate phage will lyse a fraction of the susceptible cells it infects and make the rest lysogenic; the proportion of lysogenized cells depends on temperature, metabolic conditions, and multiplicity of infection (see Lwoff 1953). The temperate phage can become preprophage in a cell already lysogenic for a related prophage. It can give rise by mutation to more virulent phage mutants, which fail more or less completely to elicit reductive responses.

The decision as to which response a cell will give to infection with a temperate phage is made very early. It requires several steps; different virulent mutants of a temperate phage, which fail to give a reductive response, can have impairments or blocks at one step or another; in mixed infection they may cooperate to elicit a reductive response (Levine 1957; Kaiser 1957). Some of these steps relate to protein synthesis, since inhibitors of protein synthesis increase the frequency of reductive responses (L. E. Bertani 1957).

One of the earliest events following DNA injection is a profound rearrangement of the nuclear apparatus of the bacterial cell. The nuclear bodies are often distorted, sometimes fragmented and peripheralized (Luria and Human 1950). With intemperate phages these changes are extensive and apparently irreversible; they may be the cause of the irrevocable cell death and of the suppression of specific cellular syntheses that accompany even abortive infection with these phages (see Cohen 1949). The cytological changes are milder and reversible in infection with temperate phages, as well as with their virulent relatives (Whitfield and Murray 1954).

In the light of the fact that prophages locate themselves within the genetic apparatus of the lysogenic cell, it seems plausible to interpret the relatively gross cytochemical changes that follow infection as the manifestation of processes tending, whether successful or not, toward a physical contact between the entering phage material and the genetic apparatus of the host.

The vegetative reproduction cycle

Problems of virus growth. Those infected bacteria that give a productive response remain intact for a characteristic *latent period.* The minimum and median latent period (Adams and Wassermann 1956) are characteristic for each phage. At the end of the latent period the bacterium lyses and releases new phage. The amount and composition of the phage yield can be determined both for mass cultures of infected cells and for individual cells.

Infected cells can be broken open by methods that do not inactivate infectious phage in order to examine intracellular phage at various times (DOERMANN 1952). These methods reveal that in vegetative reproduction *no infectious phage is present* until about the middle of the latent period; then the amount of phage increases till the time of lysis. We may ask, therefore, whether the "meaningful" steps of reproduction are performed late by the reappeared infectious phage or early by the noninfectious form. Extensive evidence of both genetic and biochemical nature supports the second alternative. Reproduction consists of synthesis and accumulation of noninfectious, vegetative phage elements, which then mature into inert, infectious particles (see DOERMANN 1953).

This situation raises a number of problems. A virus particle has a chemistry, a structure and an individuality. The vegetative, reproducing phage has a different structure. What is its chemistry, and how does it evolve into that of the final product, the virus particle? And what degree of individuality does it retain? The first question can be approached only by chemical studies, the second by genetic studies. Remarkably enough, the two approaches have led to a reasonably unified picture, however still incomplete; the unifying feature is the probable identity of phage DNA elements with the individual genetic elements that represent vegetative phage.

The chemistry of phage-infected bacteria. Upon infection with bacteriophage, some of the synthetic processes in the bacterial cells that are destined to lyse continue unchanged; others are deeply modified. On the one hand, micromolecular metabolism is generally little affected (COHEN 1949); specific requirements for substrates and for essential metabolites are the same for growth of uninfected cells and for phage production, and cell enzymes continue to function during infection. A few significant exceptions, relating mainly to nucleic acid constituents (COHEN 1953), will be mentioned below. On the other hand, macromolecular metabolism is deeply altered: synthesis of viral proteins and nucleic acid becomes a major feature.

Four methods are available to study such syntheses: 1. Determination of total amounts of a given macromolecular fraction (protein, RNA, DNA). 2. External interference with the synthesis of a specific fraction (for example, protein) and study of the effects on concomitant or subsequent syntheses of other fractions. 3. Identification and measurement of viral materials by their chemical or antigenic peculiarities. 4. Isotopic tracing of various atoms or groups of atoms through the synthetic processes.

Much work has centered on the T-even coli-phages, especially T 2, for several reasons: (a) their DNA is naturally labeled by the presence of a unique pyrimidine, hydroxymethylcytosine (HMC), in place of cytosine (WYATT and COHEN 1952); (b) information is available on the antigenic specificity of their protein components (LANNI and LANNI 1953); (c) host-specific synthesis (for example, of many enzymes) is specifically suppressed by infection with these phages (see COHEN 1949) but not with the temperate phages. This last peculiarity may be related to the deep, irreversible cytochemical changes that follow infection with these intemperate phages; both effects may reflect an incompatibility between two types of DNA, one with HMC, the other with cytosine. The activation of a normally inhibited host DNase by T 2, and not by temperate phages (see KOZLOFF 1953), may play a role in the suppression of host-specific syntheses by removing the specific DNA templates. Other factors may play a role; indeed, coli-phages such as T 5, without HMC or any other known anomaly in DNA composition, behave very much like the T-even phages.

Apart from the peculiarities relating to host-specific syntheses, the essential synthetic processes in vegetative phage reproduction can be described as follows. First, there is synthesis of a protein fraction, different from the proteins of mature phage and not destined to incorporation into it, *but required for subsequent phage DNA synthesis.* Inhibition of this early protein synthesis by chloramphenicol results in suppression of DNA synthesis (TOMIZAWA and SUNAKAWA 1956). If inhibition sets in after some protein has been formed, the rate of phage DNA synthesis parallels the amount of early protein formed. In the absence of inhibitors, specific phage DNA begins to increase after a few minutes and piles up until the time of lysis. Two to ten "phage-equivalent units" of DNA (1 unit $= 2 \times 10^{-17}$ g. DNA phosphorus) may be synthesized per minute, representing a rate of DNA synthesis higher than in uninfected cells (HERSHEY 1953). RNA synthesis is largely blocked in T 2-infected cells (see COHEN 1949), but labeling experiment reveals that there is synthesis of a small RNA fraction with a notable base composition, resembling (*mutatis mutandis*) that of phage DNA (VOLKIN and ASTRACHAN 1957).

Proteins measurable as antigens present in phage particles (a "tail antigen" and a "head antigen"; LANNI and LANNI 1953) appear towards the end of the eclipse period, one or two minutes before the appearance of the first infectious phage. At the same time, electron microscopy on disrupted cells reveals recognizable phage components, such as head membranes and tail elements (KELLENBERGER and SÉCHAUD 1957). Masses of free DNA-like material are also recognizable. Infectious phage and morphologically complete phage particles appear a few minutes later and accumulate alongside the noninfectious, phage-specific materials. Some such noninfectious materials are liberated along with phage upon lysis.

Tracer experiments have shown that the phage-type DNA synthesized in noninfectious form is a true precursor since it is transferred without breakdown to infectious phage (HERSHEY 1953). The early protein, as already mentioned, is probably not phage precursor protein in this sense. The late proteins formed before (or in excess over) the mature phage may be true precursors, but definite evidence for their incorporation into infectious particles is still lacking (HERSHEY and MELECHEN 1957). Noninfectious protein not associated with DNA as found in disrupted cells, might conceivably be "rejected" material or, more probably, a product of breakdown of phage particles that had not yet become fully stable. Genetic evidence, to be discussed later, indicates, however, that not all protein of a given phage particle is synthesized directly under the control of that particle (STREISINGER 1956). Hence, the existence of unassembled phage-precursor protein becomes a necessity.

The overall picture of phage syntheses fits the concept of multiplication of viral materials in noninfectious form. These syntheses are dependent on the metabolic machinery of the host for energy and for constituent micromolecules; the virus-specific features appear at the macromolecular level, that is, at the same level at which the specific features of the macromolecular elements of any cell must arise.

Some metabolic alterations occur at other levels. Thus, HMC (or rather, its deoxyribotide) is synthesized *de novo* in T 2-infected cells, partly by hydroxymethylation of preexistent deoxycytidilic acid (COHEN 1953). Thymine synthesis is resumed upon T 2-infection of a thymineless mutant (BARNER and COHEN 1954); pentose metabolism is shifted in the direction of desoxyribose synthesis (COHEN 1949). These changes probably reflect a variety of metabolic alterations: changes in pathways by altered demands for end-products; activations of nonfunctional host enzymes; and possibly formation of new enzymes by gene-like activity of the virus itself.

The chemical origin of phage constituents. Two questions arise in interpreting the biochemical picture. First, what are the sources for the new viral material? More specifically, do these represent *de novo* syntheses from aspecific materials or modifications of already specific macromolecular precursors? Second, in what chemical form is the essential specificity of virus, that is, its detailed genetic information, preserved during synthesis?

Chemical work provides only partial answers, which must be integrated with genetic evidence.

The question of the origin of viral constituents can be answered quite unambiguously by tracer experiments, at least for viral protein and DNA. These are both made from nonspecific precursors (with the exception of the contribution from the parental phage DNA; see below). There is no measurable destruction of cell protein during infection, nor any significant transfer of protein-specific labels from preinfection cell proteins to viral proteins. Host DNA, on the contrary, largely disappears and its nucleotides appear in phage DNA, with conversion of cytosine into HMC (for phage T 2). This host contribution to viral DNA is not a transfer of intact DNA molecules or large submolecules. Kinetic analysis of nucleotide synthesis and transfer and of competition between exogenous and endogenous nucleotides and nucleosides reveals that both the host contribution and the external contributions are merged in a pool of DNA precursors (probably consisting of single nucleotides or oligonucleotides), which feeds the DNA synthesis without preferential selection according to origin (HERSHEY 1956, 1957). After mature phage begins to be formed, there is established a steady state situation, in which DNA constituents from available sources enter the metabolic pool, are incorporated into a pool of phage-precursor DNA, which in turn is depleted at a constant rate by the maturation of phage particles (HERSHEY 1953). In the steady state, the pool of precursor DNA contains about 40–60 phage-equivalent units. A phosphorus atom takes 7 minutes on the average to reach the DNA pool and 14 minutes to reach the mature phage. It seems definite that the DNA in the phage-precursor pool is not associated with phage protein (Hershey and Melechen 1957).

Coming now to the question of the chemical form in which the genetic information is maintained and transmitted in the course of virus multiplication, most of the chemical picture presented above would seem to support the claim of DNA to be *the* genetic material, controlling its own replication and the specificities of the proteins that are ultimately found in phage. Some observations, however, weaken this claim. First, the requirement for early protein synthesis as a prerequisite to DNA synthesis and a controller of the rate of DNA synthesis (Tomizawa and Sunakawa 1956), may bespeak a compulsory transfer of information from viral DNA to a protein or to a RNA-protein complex. Second, the effects of radioactive decay of P^{32} atoms on phage multiplication (Stent 1955) are difficult to reconcile with the concept of DNA as the irreplaceable carrier of genetic specificity. Multiplication of phage T2 is prevented by decay of P^{32} incorporated in phage DNA (1 inactivation per 10 disintegrations; Hershey et al. 1951), whether the decay occurs in the free phage or immediately after infection (which can be tested by immersing the infected bacterial culture in liquid N_2, allowing time for radioactive decay, then thawing and testing for phage-producing ability). If, however, P^{32}-labeled bacteria are infected with P^{32} phage in P^{32} medium, so that all phage DNA must be equally labeled and subject to P^{32} decay, and 8—10 minutes development is allowed to occur before the cells are frozen, the phage-producing ability of the cells becomes decay-insensitive. Return to unlabeled medium after any length of time in the frozen state restores full phage production. Unless some unknown and remarkably efficient reactivation procedure takes place in these bacteria, we must assume the presence in the phage-infected cell of a decay-resistant mechanism, which has taken over the genetic information from DNA and has preserved it for later use.

What role, if any, does the small metabolically active RNA fraction play in DNA synthesis? Its base composition, resembling in base ratios that of phage DNA, may bespeak either a template or a precursor role. The whole network of interrelations between the synthesis of protein, of RNA and of DNA is at present under active investigation, both in phage synthesis and in other systems.

Additional evidence concerning the problem of continuity of phage specificity in replication derives from studies on the transfer of labeled atoms from parent to progeny phage. These will more usefully be discussed in a later section, following the description of the purely genetic approach to virus multiplication.

Genetic analysis of phage reproduction. Phage breeds true in vegetative multiplication and maturation. Two kinds of processes can give rise to progeny phage stably different from the infecting one: (a) mutation, and (b) recombination. Both processes yield information on the replication process.

Mutations in phage occur only during multiplication, not in the infectious, resting state. A bacterial cell in which a phage mutation occurs will generally yield a mixture of parent and mutant types. For rare mutations, not more than one mutation per cell will generally occur in one cycle of phage infection; thus, each mutation gives rise to one clone of mutants in the yield of a cell. The clone size distribution, observed for several types of mutations, is as predicted by assuming exponential phage reduplication and a constant probability of mutation per reduplication (LURIA 1951). Thus, we conclude that multiplication of vegetative phage is exponential, that is, that replicas of the initial phage material serve in turn as sources of replication. The process is formally like any vegetative, clonal growth process, not like the printing of successive copies from a unique mold. Hence, if genetic information is transferred to materials other than DNA these materials must also multiply (unless they simply remain available as genetic memory, not used except if DNA itself is incapacitated).

Mutations can affect any one of a variety of phage properties. Related strains differing in one, two, or more recognizable characters (plaque type, host range, requirements for attachment to cells) are easily obtained. Bacteria infected with a mixture of two such related viruses, or with a mixture of two related viruses of unknown common ancestry, such as T 2 and T 4 (DELBRÜCK and BAILEY 1946), yield mixed phage progeny. The mixture contains both parental types and types with various new combinations of the parental characters (HERSHEY and ROTMAN 1949).

This *genetic recombination* in phage "crosses" has been studied in some detail with a number of phages (T 2, T 4, T 1, λ coliphages; a *B. megaterium* phage; *Salmonella* phages) and has been observed with all phages where it has been looked for. The features of genetic recombination that are most pertinent to the problem of virus multiplication are as follows (see VISCONTI and DELBRÜCK 1953):

1. Each character or "marker" behaves as stable unit factor, transmitted to progeny phage as the Mendelian factors of classical genetics. The phage particles behave as haploid with regard to most markers.

2. In crosses between phages differing by 2, 3, or more markers the frequencies of various recombinant types fit a model of unit factors located in a linear sequence on one linkage group (possibly sometimes two groups), with frequency of recombination between factors proportional to their additive linkage distance. If the distances are expressed in percent recombination (corrected for certain peculiarities of phage genetics, see below) the known length of the T 2 genetic map is over 200 units ("centimorgans"); that of phage λ about 30 centimorgans.

3. The most distant markers (possibly unlinked) yield less than 50% recombinants in the pooled progeny from a population of infected bacteria.

4. The map distances obtained directly from recombinant frequency are not strictly additive, but exhibit an excess of multiple exchanges ("apparent negative interference").

5. For any pair of markers the recombinant frequency is lower among the first maturing progeny particles, and rises in the later, larger crop.

6. Infection with 3 suitable types gives rise to triparental recombinants.

The above facts are adequately interpreted (Visconti and Delbrück 1953) by viewing a phage "cross" as analogous not to a cross between two individuals but to a population experiment. In the infected bacterium there is a "pool" of vegetative phages, which multiply, mate (giving rise to recombinants), and are withdrawn from the pool by maturation. The quantitative analysis permits an estimate of the average number of "rounds of matings" in the ancestry of progeny phage from one intracellular growth cycle. With phages T 2 and T 4 this number increases from 2 or 3 for the earliest maturing particles, to 5 or 6 for yields of 100–300 phages, and to 7 or 8 for exceptionally late, larger yields. With other phages the number of rounds of matings is generally lower (0.5–1 for λ, Kaiser 1955; 0.5 for T 1, Bresch and Trautner 1955). Incidentally, the presence of recombinants among the earliest maturing particles is direct evidence for their being new phages, and not initial particles regaining the infectious form. Assuming a constant rate of maturation, the size of the pool of vegetative phage elements can be estimated at 30–60 elements per cell for phage T 2.

7. Reciprocal recombinant types are present in approximately equal numbers when the phage progeny from many bacteria is averaged. But in individual bacteria, reciprocal recombinants are produced in widely disparate numbers, with no significant correlation. Careful analysis (Bresch 1955) supports the conclusion that only one recombinant is produced in each elementary act of recombination (at variance with the classical process of crossing-over in meiosis or mitosis, which gives rise to two reciprocal recombinant types).

8. The numbers of particles of any one recombinant type in single bursts are not random, but show a clonal distribution. This indicates that the recombinants, once formed, can still multiply, being part of the vegetative phage pool (Stahl 1956).

9. A small, constant proportion of progeny particles, characteristic for each phage, is "heterozygous" for any one pair of parental markers; these heterozygotes segregate out progeny with either marker (Hershey and Chase 1951). The heterozygous region is invariably small, averaging 4–5 centimorgans and never being longer than 8 or 10. No complete heterozygotes have been found. Particles heterozygous for certain markers are generally recombinant for markers situated on either side of the heterozygous region.

The frequency of heterozygotes is compatible with the hypothesis that each recombination act produces a heterozygote, which then upon replication gives rise to recombinants (Levinthal 1954). Mature particles, accordingly, would be heterozygous for those regions where exchanges had occurred at the last mating and segregation had been put off by maturation until the initial replication within the next infected cell. Replications of heterozygotes as such for a few generations may also occur (Edgar 1956).

The vegetative form of phage. The genetic facts outlined above can easily be matched with the chemical data if we assume the genetic material in the vegetative phase to be phage DNA. Both phage DNA and vegetative phage accumulate, both are withdrawn irreversibly by maturation. The pool of phage T 2-precursor DNA is comparable in size to the calculated pool of vegetative phage. Various attempts, both theoretical and experimental, have recently been made to interpret mutations and recombination in phages in terms of DNA structure and synthesis and to investigate the possible pitfalls of identifying vegetative phage and DNA.

The currently accepted structure of DNA (WATSON and CRICK 1953) is that of a double helix of complementary polynucleotide chains of opposite backbone polarity, coiled around each other and hydrogen bonded to one another through pairs of bases (adenine-thymine and guanine-cytosine). This double helix carries a double set of information as base sequence. Such a structure is eminently suitable as a replicating structure since it can give rise to an identical replica not by identical copying (a chemically difficult task) but by complementary copying of polynucleotide chains. Replication of coiled DNA double helices requires an almost prodigious amount of unwinding and rewinding at each replication unless the parental polynucleotide chains break and reheal; but the energetics of the rotation required for the winding is by no means prohibitive (LEVINTHAL and CRANE 1956). However the base sequence may act to direct other specificities, such as protein structure, the amount of information in a DNA fiber (molecular weight around 6×10^6, that is, about 10^4 nucleotide pairs) is clearly sufficient for genetic purposes.

In attempting to equate DNA with vegetative phage, the replication of DNA must be made to account for the following observations: (a) the absence of correlation between reciprocal recombinants and, presumably, the formation of one recombinant at a time; (b) the presence and frequency of partial heterozygotes; (c) the transmission of material constituents from parental phage particles to progeny particles.

It is possible to account for the genetic facts by assuming that vegetative phage is DNA and that it reduplicates, as in the original WATSON and CRICK model, by the production of two new polynucleotide chains, with unwinding of the parent double helix and rewinding of the four chains to reconstitute two DNA molecules. Two reduplicating phage DNA elements can come together in homologous juxtaposition. The process of replication of one polynucleotide chain may then shift from copying one parental double helix to copying the corresponding chain of the other ("copy choice" mechanisms; LEDERBERG 1955). This shift represents the elementary recombination event. According to a model developed by LEVINTHAL (unpublished), the shift would create a strained chain complex and a second shift would then occur in the copying of the other chain. This results into the formation, as replica, of a new double helix in which one chain is "recombinant" for characters on the two sides of the first shift, and the other chain is recombinant for characters on the two sides of the second shift. The new double helix is "heterozygous" for whatever characters distinguished the two parents within the short region between the two shifts. This model can also account, by residual strains in the recombinant DNA chains past the shift points, for the observed tendency of repeated recombination to occur within short regions of the phage map ("true negative interference"; CHASE and DOERMANN 1958).

A number of findings with the effects of radiation damage on the frequency of recombination (DOERMANN et al. 1955; JACOB and WOLLMAN 1955) can be explained by the above model directly or with natural extensions, by assuming that when the process of replication reaches damaged points in the parental DNA replication is arrested, so that the "partial replicas" can only be completed by shifting to an undamaged parental template.

Transfer of parental phage DNA to the phage progeny. The transmission of labeled atoms from parent to progeny phage takes place in a way consistent with the idea that DNA replication, as outlined above, is the essential feature of phage reduplication. P^{32} atoms (or labeled bases) from infecting phage are transferred to progeny phage with an efficiency of 50% or more at each growth cycle (MAALØE and WATSON 1951). Sulfur atoms or amino acids from the protein of infecting phage are not transferred (HERSHEY and CHASE 1952). The distribution of the parental P atoms among the progeny particles is informative as to the mechanism of replication. If the DNA is the genetic material and is replicated without breakage of the polynucleotide chains, then the parental label should be concentrated in one progeny particle (fully "conservative" replication) or in two progeny particles ("semiconservative" replication) depending on whether the two polynucleotide chains of the initial particle remain together or separate. If the parental chains are broken up and reshuffled with fragments of the new ones, then we should observe "dispersive" replication (DELBRÜCK and STENT 1957). The DNA theory of genetic recombination outlined above predicts semiconservative replication of the genetic material. Experiments in which the P^{32} atoms of phage T 2 were traced in the progeny phage by autoradiography for several successive growth cycles (LEVINTHAL 1956; LEVINTHAL and THOMAS 1957) showed that for each initial phage particle there appear in the progeny phage not more than two fragments, each containing about 20% of the P atoms of an infecting particle. Examination of the DNA released by breaking open the labeled free phage reveals a single large fragment per particle, containing 40% of the phage DNA. This fragment can also be distinguished by peculiar chemical features (BROWN and SYMONDS 1957). The large DNA fragment is transmitted semiconservatively to the progeny. The rest of the DNA is presumably in small molecules, whose fate in replication is unknown.

An alternative method of following the distribution of parental P^{32} atoms in the progeny phage, by the loss of infectivity due to radioactive decay, gives results similar to those of autoradiography (DELBRÜCK and STENT 1957). Finally, in mixed infection with a labeled and an unlabeled parent, there is a persistent association of parental atoms with genetic characters derived from the labeled parent, as predicted by the semiconservative replication hypothesis (LEVINTHAL and THOMAS 1957; HERSHEY and BURGI 1956).

The above picture of phage reproduction as consisting of the replication of coded DNA molecules, which can mate by homologous pairing while being replicated, satisfies the peculiarities of phage genetics, such as the absence of reciprocal recombinants in the act of recombination, the presence of short heterozygous regions, and the localized negative interference. A paramount task will be to reconcile this mechanism of recombination with the mechanisms of classical chromosomal pairing and crossing-over, which, instead, give rise to reciprocal recombinants.

Phenomena resembling those of phage genetics are observed in yeast, *Neurospora* and other organisms, which usually exhibit classical chromosomal mechanisms, when recombination is studied in certain small genetic regions (LINDEGREN 1955, ROMAN 1956).

A most important recent finding is that the transmission of tritium-labeled DNA in mitotic reduplication of chromosomes of plant cells follows the same semiconservative pattern as the transmission of DNA in phage replication (TAYLOR et al. 1956). This result has led to the suggestion of a unifying scheme (TAYLOR 1957), in which a chromosome is viewed as a dual structure, with a linear backbone, exhibiting classical "four strand" crossing-over, and a set of DNA side-chains, exhibiting phage-like, copy-choice recombination. This stimulating hypothesis, which predicts that branched genetic maps will be observed in suitable tests, should be amenable to direct experimental verification.

DNA and the fine genetic structure of phage. An unusual experimental system has provided a remarkable insight into the fine genetic structure of phage (BENZER 1955). Any one of a number of independent *r* mutations in phage T4 suppress the same property, namely, the ability of the phage to multiply in a certain host. All these mutations affect a set of genetic sites that are clustered in two adjacent short regions of the phage linkage map, as shown by tests of recombination between any two *r* mutants. Since only r^+ recombinants can grow on the special host, this system permits the measurement of very low recombination frequencies. The results show that there is a minimum recombination distance at which two mutational sites can be located: the recombination frequencies range down to about 0.01% then drop to less than 10^{-6} (probably 0). This suggests a "quantum" of recombination of the order of 10^{-4}. Assuming, (a) that the genetic material is DNA; (b) that 40% of all the DNA of a phage is indispensable genetic material (as suggested by the P^{32} transfer experiments); (c) that genetic recombination represents exchanges or copy-choice shifts between nucleotides; and (d) that the frequency of recombination is uniform per unit length of DNA, this analysis leads to the remarkable conclusion that recombination can occur between genetic elements not longer than a few nucleotides and, more probably, between single nucleotides along the polynucleotide chain. Besides estimating the quantal unit of recombination (or "recon"), this analysis permits an estimate of the size of the genetic element affected by mutation (or "muton"), which may vary from few (and possibly one) to hundreds of nucleotide pairs, and of the unit of genetic function (or "cistron" within which any one mutation gives the same effect), which may cover several thousands of pairs (BENZER 1957).

Other genetic systems in phage confirm and extend this fine structure analysis (Streisinger and Franklin 1956).

Phage replication, DNA replication, and transfer of genetic information. The assumption that the vegetative phage is DNA has led to a completely plausible model of the multiplication and genetic organization of phage. Yet, it is by no means certain that DNA is *the only* reproductive form in which the genetic information of phage can be embodied, nor is it excluded that non-DNA elements play a regular role in reproduction. Several lines of evidence already mentioned suggest such a role. We have, first, the need for an early protein synthesis to control phage DNA synthesis; second, the stability to P^{32} decay of the phage-producing ability of an infected cell once a critical stage has been reached; and third, the production of some apparently specific RNA as part of phage replication.

To account for these findings in a unitary way, an ingenious theory has been proposed (Stent 1957), which postulates that specific RNA-protein replicas are necessary intermediates in DNA reproduction. First, a DNA molecule (double helix) would give rise to a single-fiber RNA-protein replica. Then, two such RNA-protein templates would come together in homologous pairing and together control the synthesis of a new DNA double helix. If the two paired RNA-protein replicas are homologous but not genetically identical (that is, if they carry allelic differences) there would be created an opportunity for copy-choice shifts and recombination.

This theory would make mating (between two RNA-protein replicas) an essential feature in the reproduction of DNA double helices. This would account for the remarkable similarity between number of phage generations and number of rounds of matings in the multiplication of phages T2 and T4.

A daring extension of the theory suggests that whenever the number of rounds of matings, as calculated from recombination data, is smaller than the number of phage generations, as with phages T1, λ and many others, mating might occur, not only among RNA-protein replicas of the phage itself, but among replicas of the phage and replicas of phage-homologous regions of the host-cell DNA. This suggestion draws its basis from the phenomenon of lysogeny, in which the prophage behaves as a group of chromosomal elements of the host cell (see below). It could account for a number of puzzling observations, where a host cell appears to repair radiation damage to phage DNA and to induce specific mutations in the phage that multiplies within it (Weigle 1953, Garen and Zinder 1955).

The validity of this ingenious theory, with its many ramifications and *ad hoc* hypotheses, will have to be judged by the chemical plausibility of the molecular structures it postulates, by the possibility of demonstrating the specific intermediates predicted by the theory, and by the verifiability of whatever specific predictions it may make about events in phage growth and genetics. An important objection to the theory, namely, that good phage-precursor DNA can be made in the presence of chloramphenicol (Hershey and Melechen 1957) would lose its validity if chloramphenicol were to allow continued synthesis of an RNA-associated protein (SPIEGELMAN, personal communication).

Mutations in phage by chemical treatments have only been obtained when phage was made to incorporate 5-bromuracil instead of thymine in its DNA (LITMAN and PARDEE, 1956). It would be of great interest to test if phage mutations could also be induced by treatments that can specifically modify the composition of RNA.

Phage maturation, phenotypic mixing, and cell lysis. Little is known of the chemical events in the terminal phases of phage development. Maturation can be prevented by chemicals such as proflavine, which permit accumulation of all the known phage constituents (DEMARS 1955), but prevent the formation of mature infectious particles. Empty membranes of phage head, obtained from cells lysed in proflavine, reveal an antigen distinct from those of the phage tail. The latter antigens can be identified with the cell-combining sites, which determine the host range of the phage, and are determined by a number of genetic factors (STREISINGER 1956 a). In mixed infection with related phages, there is observed *phenotypic mixing* of tail antigens. For example, a bacterium infected with two types of particles, with tail properties A and B, can produce particles that are phenotypically A (as judged from their cell-combining and antigenic properties) but genetically B (as judged from the stable properties of their progeny) and vice versa. Some particles are phenotypically A *and* B (STREISINGER 1956 b). Phenotypic mixing occurs independently for at least two different properties of the phage tail: host range and cofactor requirement for cell attachment (BRENNER 1957).

These observations show that the proteins incorporated into a mature phage particle are not formed directly under the direction of the genetic material of that very particle, but have been produced under the control of any one or more of the phages present in the cell and are used more or less at random in the assembly of mature phage.

Little is known of the mechanism of lysis. Premature lysis can be induced by treatments such as chloroform or streptomycin (SÉCHAUD and KELLENBERGER 1956), but only after the process of maturation (or, at least, phage-protein synthesis) has started. It can sometimes be delayed by superinfection with related phages. The relation, if any, between lysis at the end of the latent period and damage to the cell surface by phage attachment (see TOLMACH 1957) remains obscure.

Lysogeny

We shall now consider the host-phage interaction in the "reductive response" leading to lysogeny (see LWOFF 1953; BERTANI 1958). The infected cell survives and, often after a short delay in growth, multiplies. Its early progeny consists of a mixture of lysogenic and nonlysogenic, phage-sensitive cells (LIEB 1953). Stably lysogenic segregants arise by what appear to be individual acts of prophage establishment, of which there can be many within the progeny of a single infected cell (FRASER and LURIA unpublished). Some phage mutations interfere with the ability to accomplish successfully this act of fixation (LEVINE 1957; KAISER 1957).

The stably lysogenic cells give rise to very few, if any, nonlysogenic descendants. They are usually "immune" to lysis by reinfection with the same phage or its relatives; immunity is itself subject to control by specific phage genes (BERTANI 1957).

Lysogeny reveals itself by the occasional production of mature phage. The proportion of lysogenic cells that produce phage in each cell generation is specific for each system. It may vary from 10^{-2} to 10^{-6} or less; it may be increased (often up to 100%) by certain "inducing" agents such as ultraviolet light (see LWOFF 1953). Production of infectious phage is accompanied by lysis and is preceded by a process analogous to the vegetative multiplication of the same phage in cells that give a productive response to infection. Lysogenic bacteria derived by infection are indistinguishable in all respects from lysogenic bacteria isolated from natural environments. Indeed, most strains of bacteria are lysogenic and it is clear that prophage is the form in which temperate phages are most commonly perpetuated in nature. A bacterium lysogenic for several unrelated phages will usualy liberate only one of them in one lysing event.

The frequency of phage production by lysogenic cells is genetically controlled. *Defective* prophages may lose altogether the ability to give rise to infectious phage (JACOB and WOLLMAN 1956). Their presence may be revealed either by the inducibility of lysis (without production of infectious phage), or by persistent immunity, or by occasional mutations to a nondefective form.

Defective prophages, therefore, can control immunity and other host properties (see below) without manifesting themselves as progenitors of infectious phage. Since by prophage we mean both the property of a cell to be a potential phage producer and the material element carrying this property in the lysogenic cell, defective prophages raise the question of whether we can meaningfully distinguish a prophage from other determinants of cell heredity.

Several lines of evidence bear on this question, which is fundamental for our understanding of the relation of viruses to cells. They concern: (a) the number and location of prophages in the cell; (b) the control of cell properties by prophages; (c) the role of infectious phages as vectors for cell-to-cell transfer of genetic materials. We shall discuss these three items in the above order.

The location of prophages is revealed by crosses among strains of *Escherichia coli* lysogenic or nonlysogenic for phages of the lambda group (LEDERBERG and LEDERBERG 1953) and a few others. The results verify the hypothesis that lysogeny for a given phage is controlled by a complex genetic determinant, which is attached to the "chromosome" of the host cell, embodies all the details of the genetic specificity of the phage, and is presumably the prophage itself. The evidence for this statement (see JACOB and WOLLMAN 1957) may be summed up as follows: (a) Lysogeny and nonlysogeny segregate as unit factors in crosses between lysogenic and nonlysogenic *E. coli* strains. (b) These factors are located at specific sites in the genetic map of the host. (c) In crosses between bacteria lysogenic for different mutants of the same phage, the prophages segregate in the progeny as unit elements carrying the genetic characteristics of the parental prophages or predictable recombinants; thus, the chromosomal elements are not simply factors controlling the persistence of a prophage, but factors embodying the details of prophage heredity. (d) In transduction (see below), where a phage acts as carrier for transfer of small fragments of the bacterial chromosomes, one prophage can be transduced by a different, unrelated phage just like any other chromosomal segment, in association with adjacent genetic determinants. (e) In bacteria doubly lysogenic for two closely related prophages, the two may be located at different, distant chromosomal locations. (f) The number of copies of a prophage per cell is similar to the number of cell nuclei (Bertani 1954).

Thus, we come to consider as prophage a certain region of the bacterial chromosome that controls the potentiality of producing phage according to genetic specifications exactly embodied in the prophage itself. It is tempting to imagine the prophage as consisting of a DNA element similar to a vegetative phage embedded in the host chromosome; isotopic evidence that the prophage contains as much essential phosphorus as mature phage has been obtained (Stent et al. 1957).

Results of bacterial crosses and transduction are not yet adequate to decide whether in the prophage the phage genetic map is linearly continuous with the map of adjacent host factors; nor do we know yet whether lysogenization constitutes an insertion or a replacement of genetic material, that is, whether the nonlysogenic bacterium possesses "nonprophage" regions homologous to the prophages and replaceable by the prophages. The apparent genetic homologies between host cells and temperate phages, suggested by several indirect lines of evidence discussed in a preceding section, tend to support at least a partial replacement theory, probably by a copy-choise replication mechanism. These intriguing aspects of phage-host relationship are under active investigation; one of their most exciting prospects is the possibility of unifying the genetic mechanisms of the phage and of the bacterial cell.

The cell properties controlled by prophages are not limited to immunity and inducibility. They include properties which are seemingly unrelated to the life cycle of a virus. The two most remarkable cases are the control of diphtheria toxin production in *Corynebacterium diphtheriae* (see Groman 1955) and the control of some somatic antigens in *Salmonella* (Uetake et al. 1955). In each of these cases every cell that becomes lysogenic for a certain prophage acquires the new property. These cases of *conversion* by lysogenization are only the most impressive examples of genetic roles of prophage; others include control over phage susceptibility, colony types, and metabolic patterns (see Bertani 1957 b).

Indeed it is conceivable that all prophages control some function of which we may be unaware. Conversely, it is conceivable that any genetic determinant of bacteria may be or become part of a prophage. It is interesting to recall that even in productive infection, followed directly by vegetative multiplication, there is evidence for control of cell functions by the genetic material of phage. Thus, in infection with T2, cells are forced to synthesize the new nucleotide of HMC, and a thymineless mutant of *E. coli* infected with T2 acquires the ability to synthesize thymidilic acid (Cohen 1953). Phages of *Klebsiella* and other capsulated organisms determine, during vegetative multiplication, the production of capsule-digesting enzymes, which may even become incorporated into the protein of mature virus particles (Adams and Park 1956). In *Salmonella,* a phage-controlled somatic antigen can be demonstrated within 10 minutes after infection with a virulent mutant of the controlling phage, in cells that will all lyse within another half-hour (Uetake et al. 1958).

Thus, vegetative phage, as well as prophage, can function genetically in controlling synthetic processes at various levels, although in vegetative reproduction most new syntheses appear to be directed to the production of phage materials. In this light, phage maturation can be viewed as the inclusion of portions of genetic material into specific structures, the infectious particles, which are necessary vehicles for the transmission to other bacterial cells of these bits of "infectious heredity."

Transduction. It is not surprising, therefore, that genetic elements of the cell, other than those that we call "phage" because they are capable of vegetative multiplication leading to lysis, can gain transit aboard an infectious phage particle. The phenomenon of transduction consists of the transmission to a "recipient" cell, by infection with a temperate phage, of some distinctive character of the bacterial cells in which the phage has grown last ("donor" cell) (Zinder and Lederberg 1952). This event is generally rare (1 in 10^6 cells among the survivors of infection). Only one host character or a small group of closely linked characters is transmitted at a time (Lennox 1955). The frequency of joint transduction drops rapidly with map distance; the maximum fragment size is probably of the order of $^1/_{50}$ of the cell genome. If we equate DNA with genetic materials, this amount of host DNA is comparable to the DNA content of one phage particle.

When all characters of the donor have equal chance to be transmitted, it seems likely that more-or-less random fragments of host chromosome become transient passengers aboard the maturing phage particles (Zinder 1955). One unusual instance of specialized transduction (Morse et al. 1956 a, b), that of the galactose determinant by phage λ, seems to reflect a more intimate, lasting association between vegetative phage and a particular fragment of cell genome of which it has once been a chromosomal neighbor in the prophage state (Arber et al. 1957). The galactose controlling chromosomal segment is apparently incorporated in the phage genome and replaces some of the genetic material of the phage, which becomes thereby defective.

Modifications of phage by the host. Some phenomena observed with phage suggest an interaction between the genetic materials of phage and host possibly less profound than those leading to permanent recombination.

Typical is a situation where most of the particles of a phage, after growing in a given host X, are unable to multiply in a different host Y. The few that can multiply in Y, however, give a progeny fully adapted to Y; but a single cycle of growth in X will completely remove the adaptation to Y. More complex interactions are also encountered (see LURIA 1953). These reversible, nonhereditary modifications affect the ability of the phage material to initiate the process of reproduction. They reveal a mechanism by which adaptation of viruses to new hosts in nature may take place in the absence of mutation and selection.

In conclusion, this review of the present status of our knowledge on phage reproduction has led us to the concept of phages as genetically complex fragments of bacterial heredity, capable of reproduction either as prophage, in synchrony and integration with the bacterial genome, or as vegetative phage, leading in unchecked multiplication to the destruction of the bacterium. The latter process creates the viral forms, or infectious particles, capable of performing the role of vectors of genetic fragments from cell to cell. How widespread the potentiality for viral transmission is among various portions of the bacterial genome, and possibly even in cells other than bacteria, remains to be decided.

Multiplication of Viruses in Cells of Vertebrate Hosts

Properties of the viruses

The analysis of multiplication of viruses pathogenic for vertebrates has progressed less far than that of phage. This is due to a number of reasons.

In the first place, a detailed picture of virus multiplication can only emerge when all possible approaches are followed and integrated with one another; this has not yet been possible for any one group of animal viruses. Thus, present knowledge rests mainly on kinetic (growth curves), morphological (electron microscopy and cytochemistry), and antigenic analysis. Biochemical, isotopic, genetic and radiobiological approaches, which have contributed greatly to a unified picture of phage production, are only beginning to be applied to animal virus work. Studies of chemotherapy have made contributions to the definition of certain phases of virus growth, without clarifying its mechanisms. The difficulties reside mainly in the relative complexity of the host systems and are rapidly being removed by progress in the study of cultivated animal cells as microorganisms (PUCK 1957). Just as phage work flourished in parallel with the tremendous development of bacterial biochemistry and genetics, thus will animal virus work interplay with research on the mammalian cell.

In the second place, animal viruses and their hosts include a tremendous variety of systems, which differ from one another in nature, relation to host cells and metabolic requirements, much more than any phages do. Practical considerations have scattered research efforts among many different viruses, some of them poorly qualified to serve as prototypes, such as influenza viruses, which are handicapped by thermal instability, or rabbit papilloma virus, for which only inefficient titration methods are available. Isotopic studies on virus multiplications are greatly handicapped by the presence of many inactive virus particles and by low "infectivity : particles" ratios, which preclude a decision as to whether the observed results reflect the multiplying virus or the inactive or unsuccessful material.

The variety of animal viruses has made some workers doubt the value of the model approach, both within this group and with regard to extrapolation from phage research. Clearly, phage work can contribute both methodologically—because of the intrinsic operational similarities of all studies on cellular infection—and biologically—because of the insight it provides into the relation of exogenous constituents to cell organization, function and heredity. That working hypotheses derived from phage work should not become dogmas in other areas of virology is a truism worth reiterating, but not a ground for adopting a self-defeating ecclecticism. As for model animal viruses, their choice is dictated by the major virus types and by their individual properties. Chemistry, rather than size or shape of virus particles, is probably a reliable guide. Just as presence of DNA as the only nucleic acid in all phages is a convenient and probably safe unifying postulate, thus mammalian viruses containing only RNA or only DNA, or possibly a mixture of both in their infectious particles, may constitute natural categories, not necessarily (or even probably) in a taxonomic sense, but with regard to the possible community of biochemical requirements, genetic function, and relationship to cell constituents. This viewpoint is also supported by the similarity in the amount of RNA present in the particle of such disparate viruses as poliomyelitis, influenza, and several plant viruses (Frisch-Niggemeyer 1956). Our present ignorance of the chemical composition of any but a handful of viruses need not preclude some such simplifying assumption as a guide to choice of material.

The RNA viruses include the so-called MNI group (mumps-Newcastle-influenza, plus fowl plague); the poliomyelitis group; probably the equine encephalomyelitis group. DNA viruses include the vaccinia group, papilloma and possibly herpes and the APC viruses. No fully convincing proof that virus particles can contain both nucleic acids is available, although the psittacosis group may (see Moulder 1954). In addition, the RNA viruses can be subdivided into two major groups with relation to sensitivity to ether (Andrewes and Herstmann 1949): one group, ether resistant, contains only RNA and protein (model: polio); the other, ether sensitive, contains also lipids (model: MNI). The protein moiety of the polio virus may be a single protein; that of the MNI viruses includes a multiplicity of antigens, some of which cross-react with those of uninfected host cells (Knight 1946, Smith et al. 1955). Vaccinia is a prototype of viruses with DNA and multiple proteins (Smadel and Hoagland 1942). DNA viruses may exist that contain only one protein type.

Faced with such a variety of partially investigated systems, it seems wise to approach a description of virus multiplication from the standpoint of one group, that of the MNI viruses, and to comment on other viruses in connection with phases where their behavior has revealed significant differences or confirmations. Within the MNI group, the influenza viruses have been studied most extensively (see HENLE 1953; BURNET 1955), but their limited activity on available tissue culture systems and their thermal instability is pushing to the fore the more stable Newcastle-disease virus (NDV) and fowl plague virus (FPV).

The infectious particles of the viruses of the MNI group (including FPV) have the following characteristics, which are relevant to an understanding of their multiplication. Most particles are spherical, with average diameters of 60–80 mμ as measured in sectioned preparations (MORGAN et al. 1956) and 80–110 mμ in dried preparations of pure virus (probably flattened; WILLIAMS 1953). The particle size is variable, with a dispersion of about 20% around the mean value. The infectious materials contain about 0.7–1% RNA (probably 3–4% for FPV), 30% lipids, about 3.5% carbohydrate, the remaining components being proteins (ADA 1957; FRISCH-NIGGEMEYER and HOYLE 1956; SCHÄFER 1957). The particles have the property of hemagglutination (HIRST 1941); that is, they combine with red blood cells (RBC) and clump them together by forming bridges between them. This reaction can take place in the cold. These viruses also possess an enzyme that can attack the prosthetic group of a variety of mucoprotein substrates, including the virus-receptor substances of RBC and of host cells, which are certainly mucoproteins in nature (see BURNET 1951). A similar "receptor-destroying enzyme" (RDE) is available as a bacterial product (BURNET and STONE 1947). The site on the substrate that is attacked by the enzyme contains neuraminic acid (GOTTSCHALK 1956, 1957), which is split off the polysaccharide moiety of the mucoprotein. The enzyme permits the virus to "elute" from RBC if temperature and medium are suitable.

Pure virus particles have a major antigen, the V antigen, which embodies the type, group, and strain specificities of the virus particles. When particles of MNI viruses are disintegrated either by ether treatment or by sonic vibration (HOYLE 1952, LIEF and HENLE 1956, SCHÄFER 1957) they liberate several well defined fractions: the S antigen, which is a nucleoprotein containing all the viral RNA and with a characteristic antigen distinct from the V antigen; and the HA particles, with hemagglutination activity and V antigen, containing protein and carbohydrate but no RNA. The S antigen is 15–25 mμ in size and appears in electron micrographs as elongated, possibly aggregated particles; the HA has round particles about 30–35 mμ. Virus disintegration increases the total hemagglutinin activity, probably by freeing a number of HA elements per particle. It reveals the S antigen, which apparently lies hidden internally. Various estimates give 70 particles of S and 70 particles of HA per influenza particle (FRISCH-NIGGEMEYER and HOYLE 1956; probably too high) and 15–20 particles of S and 6 particles of HA for FPV (SCHÄFER 1957). Virus disintegration also liberates a lipid-containing fraction, which includes complex carbohydrates and protein. These cross-react serologically with host cell antigens and may be native or modified elements of the cell surface that have become incorporated into virus particles, as will be discussed later.

The virus-infected cells produce noninfectious virus-specific materials, besides the virus particles (see HENLE 1953). Free S antigen is produced and liberated in excess. Noninfectious HA elements are found both by extracting infected cells and as products of incomplete abortive infections (see below). They are frequently present not as small particles, like those extracted from the virus, but as large particles with low RNA content (SCHLESINGER 1953).

In addition, virus-infected cells produce long filaments, about 60–70 mμ in diameter (that is, smaller than the virus particles in cross-section), which have HA activity and V antigenicity but relatively low infectivity, if any (DONALDS and ISAACS 1954). Disintegration of the filaments yields increased HA activity but no extra infectivity.

All of these viruses-specific materials can be measured rather precisely. The antigens, S and V, are detected by suitable antisera by means of the complement fixation reaction. HA activity is measured under standard conditions; with the most highly infectious preparations of influenza, 1 HA unit corresponds to about 10^6 infectious units. Particles with HA activity, including the infectious virus, are easily purified by combining centrifugation with cycles of adsorption-elution from RBC; the degree of purification, however, is probably not as great as has been achieved with some other viruses.

Infectivity titration is generally done for influenza virus by end-point methods in chick embryos; NDV and FPV are more conveniently assayed by counts of lesions on tissue culture monolayers under a nutrient agar layer, where virus spreading and diffusing to contiguous cells produces localized "plaques" of cell death (DULBECCO 1952). Each infectious unit is defined as a plaque-forming unit or *pfu*. Some influenza strains, such as the neurotropic variant of the WS strain of influenza A, can also be assayed by plaque counts.

Plaque count methods are available for many other viruses, including poliovirus, equine encephalomyelitis, vaccinia and herpes simplex. Pock counts on the chlorioallantoic membrane (CAM) of the chick embryo (BEVERIDGE and BURNET 1946) provide good titration for several viruses, including non-destructive, tumor-producing viruses like that of Rous sarcoma (RUBIN 1955).

One infectious unit, as defined by titration, may contain as few as 1–4 of the characteristic virus particles, which can be counted by electron microscopy (ISAACS 1957) or, for MNI viruses, by their ability to act as bridges between red blood cells (LEVINE et al. 1953). High ratios "infectivity/particles" are observed with some preparations of vaccinia and of NDV. With influenza the best preparations may have ratios 1 : 10. With other viruses, one infectious unit may correspond to several hundreds, thousands, or even millions of particles, as with papilloma virus. These low ratios may reflect either the presence of intrinsically inactive particles or the inefficiency of the test system for infectivity; more often, probably, a combination of both. This raises the interesting question as to the chemical and biological uniformity of the "native" particles of a given virus, a question to which we shall return later.

General course of animal virus growth

There are many studies of virus growth in tissues and organs of intact animals. The virus may either remain localized near the point of entry or spread elsewhere. Differences in virus susceptibility among cells of the same organisms, apart from differences in accessibility, reflect complex interplays of developmental and regulatory processes (see Burnet 1955). Thus, for example, poliovirus fails to multiply when injected in the kidneys of monkeys but grows when the kidney cells are cultivated *in vitro* (Kaplan 1955). Clearly, intact animals are too complex for basic studies on virus multiplication.

Likewise, extensive work on the effects of nutrition of whole animals on the course of viral infections (Clark 1949) and on biochemical changes in organs infected with viruses (see Bauer 1953) has provided little insight on virus growth. Most effects that are observed reflect at least as much the secondary manifestations of cell damage as the primary interaction between growing virus and host cell. Viral multiplication may be as extensive in animals that die as in animals that show little or no signs of disease, or in animals protected by certain antimetabolites (Haas et al. 1957). Our discussion will consider mainly virus growth in simpler systems, such as in the cell layers of the membranes around the chick embryo, in monolayer tissue cultures, or in single cell suspensions. These can be obtained by trypsinization either from minced animal tissues or from tissue cultures. The latter method makes it possible to work with suspensions of pure cell lines (Gye and Bang 1951; Puck et al. 1956).

The typical course of virus infection in susceptible cells, as revealed by infectivity measurement, for example, of influenza virus infecting the allantoic endothelium of the chick embryo, includes an initial virus attachment, followed by a latent period in which no release of infectious virus takes place, and a period of virus liberation. This cycle of events, which can be observed with a whole variety of viruses in appropriate host cell systems (see Sanders 1957) merely reflects the intracellular location of virus reproduction.

If the initial period of infection is limited in time, as by washing away unadsorbed virus or by treatment with virus-neutralizing antiserum, and if recycling is precluded by preventing reattachment of the newly liberated virus—for example, for MNI viruses by treatment of the cells with receptor-destroying enzyme—we observe a single growth cycle, which permits analysis of virus production by single cells (Henle 1953). Single cycle growth is best studied with cells that can be freed from monolayer cultures, either before or after infection. These are diluted in nutrient medium to such an extent that readsorption of virus freed at the end of the latent period is prevented, as in the classic one-step growth experiment with phage. This has been done with NDV, WEE and polio viruses (DULBECCO and VOGT 1953, 1954, LEVINE and SAGIK 1956).

With suspensions infected with Western equine encephalomyelitis (WEE) or with poliovirus, release of virus from individual cells has been studied by distributing the cells into individual tubes or into microdrops and titrating samples of the surrounding fluid at intervals (DULBECCO and VOGT 1953, LWOFF et al. 1955).

The average amount of virus liberated per cell may be as high as 1000 infectious units or more; giant cells resulting from irradiation may produce 100 times more virus (PUCK and MARCUS 1956). Experiments both with deembryonated eggs (CAIRNS and MASON 1953) and with suspended cells (DULBECCO and VOGT 1953) show that virus release is not sudden, as in the bursting of a phage-loaded bacterium, but is gradual and may continue for many hours.

In some instances, such as MNI viruses, WEE and poliovirus, most of the new virus produced in suspensions of cells is probably released into the medium. With other viruses, such as vaccinia, only a small part of the new virus is released; the rest apparently is retained within the cells (OVERMAN and TAMM 1957). If a virus multiplies in a fixed cell layer, liberation may occur at surfaces other than the free ones; thus, virus can spread directly and preferentially to neighboring cells. There is evidence for this at least in instances where a focal infection develops around the initially infected cells (WILDY 1954). For these reasons, the virus released free often represents only part of the total virus produced. Cell-to-cell spread does not occur with cell suspensions; here, however, the amount of virus produced per cell is lower than in compact layers, probably because of an unfavorable metabolic state of the isolated cells. For example, suspensions of fibroblasts infected with NDV give an average yield of only 20—30 infectious units per cell, instead of several hundreds as produced by cells still on glass (LEVINE and SAGIK 1956).

If a cubical cell, 10 μ in size, produces 10^3 virus particles 100 mμ in diameter, these will total less than $^1/_{1000}$ of the cell volume. For 10^5 particles 30 mμ in diameter, the total would still be less than 1%. Such calculations provide rough indications of the order of magnitude of the synthetic demands created by virus infection.

Stages in virus-cell interaction

Attachment and penetration. The initial attachment of MNI viruses to host cells, like that to red blood cells, is mediated by receptors (mucopolysaccharides) that can be destroyed by virus enzyme or by RDE. It is generally accepted that enzyme-receptor combination is the mechanism of attachment (see BURNET 1955, TOLMACH 1957). The integrity of the enzymatic site of the virus is apparently not essential, however, since virus whose enzyme has been inactivated can still be adsorbed. Hence, the attachment needed for virus adsorption may have less strict steric requirements than for enzyme action, or may involve other molecular groupings. Attachment receptors for viruses of other groups have not been identified.

Adsorption, as would be expected, follows the kinetics of a first order reaction. The collision efficiency, that is, the frequency of collisions resulting in virus adsorption, is controlled by the nature of cell and virus and by the ionic environment (see TOLMACH 1957).

The results of titration indicate that a single particle is usually capable of initiating cellular infection. With host cells such as the human HeLa strain, which can be plated to give individual colonies, it can be shown that a single particle of NDV is sufficient to kill a cell (P. I. MARCUS, personal communication).

Virus attachment is generally not reversible by simple dilution. In competent host cells, it is followed by a process of integration leading to virus multiplication. As with phage, the use of altered virus and other methods permit the identification of a number of steps.

Ultraviolet-inactivated NDV attaches to host cells such as lung cells of chick embroys or HeLa cells, but does not kill them (P. I. MARCUS, personal communication). At least in the lung cells it can, however, exclude subsequently added active particles of NDV and prevent their multiplication. The excluded virus is apparently destroyed. The exclusion is established rapidly, within a few minutes and, most significantly, can for a while be reversed by anti-viral serum. This proves that the excluding element is still available at the cell surface and that the exclusion probably takes place at that surface. After several hours the cells with UV virus recover sensitivity to reinfection. Note that UV-inactivated poliovirus does not exclude active virus (BALUDA 1957).

Penetration follows attachment of active virus. It can be studied with NDV by the acquisition of resistance to neutralizing antiserum and to removal by RDE. Cell-adsorbed NDV becomes serum-resistant very rapidly, within 5 minutes at 37° C., much more slowly at lower temperatures, indicating a chemical transformation rather than a simple diffusion process (RUBIN, personal communication). If serum action has not gone too far, serum-neutralized virus can still attach to cells, but remains removable by RDE (RUBIN and FRANKLIN, 1957). The first order kinetics of virus neutralization indicates that one antibody molecule suffices to prevent penetration. Instead, cooperation among 2 or 3 molecules seems needed to prevent attachment. Penetration apparently requires a certain function or site of the virus, which may be blocked by virus antibody either before or immediately after attachment to the cell. The kinetics of neutralization does not imply that only one active site per particle exists; only, that a particle can only "try" one site. That is, if a particle had several sites, some blocked, some unblocked, and if attachment by means of a blocked site were irreversible, the same kinetics would obtain.

The role of virus enzyme in penetration is not clear; failure of serum-neutralized NDV to elute from RBC suggests that infectivity and enzyme activity are neutralized at similar rates. With influenza in minced CAM tissue the inhibitor d-amino-2-*p*-methoxyphenylmethane sulfonic acid

(AMPS) arrests virus infection at a stage where the virus remains removable by RDE (see ACKERMANN and FRANCIS 1954).

The actual site of the attachment reaction and of the following ones, in relation to the mucinous layers present at the surface of the host cells and to substrates that are part of the cell membrane itself, requires further clarification. The presumed active role of the cell in virus uptake has been designed as "viropexis" (FAZEKAS 1948); its function in normal penetration has been questioned (RUBIN and FRANKLIN 1957).

With poliovirus and human HeLa cells there is also a stage-wise process, in which attachment, stability to washing and insensitivity to neutralizing serum are reached in succession (ACKERMANN and PAYNE 1957). Virus that has already reacted with antiserum, but is still infectious, may remain serum sensitive for a longer time after attachment to cells. This suggest a delayed penetration (MANDEL, personal communication).

Integration, eclipse and multiplication. Following penetration, the virus materials must become functionally integrated with the cell machinery to reproduce and give rise to new virus. This process may be characterized by changes in cell function, leading to recognizable cell destruction ("cytopathogenicity"); by changes in intracellular morphology reflecting the production of virus constituents or other abnormal materials ("inclusions"); and by changes in the organization of the infecting virus itself. The initiation of these reactions may vary greatly in time among individual cells, leading to asynchrony in the process (CAIRNS 1957).

With most viruses there has been reported an "eclipse" of infectivity, during which no infectious virus can be recovered if cells are extracted by methods that do not inactivate infectious virus (see HENLE 1953, ANDERSON 1954, RUBIN et al. 1955). This eclipse need not reflect a separation of virus constituents at the cell surface, as with phage; there is a suggestion of actual penetration of the virus particle as a whole, at least with vaccina, where characteristic particles have been observed to persist within infected cells for some time after infection (BLAND and ROBINOW 1939). Eclipse would then reflect an intracellular alteration, of the infecting virus particle.

Yet, the evidence for eclipse of animal viruses (interpreted as the transformation of the infective particles into reproductive, noninfectious forms) is not completely cogent. It rests mainly on the assumption that we could recover infectious virus if it were present or, more precisely, that undestroyed virus particles, when extracted from the cell, would titrate at all times with the same efficiency as the virus in the inoculum; this need not be so. Also, the new virus might stem from the multiplication of a small amount of infectious virus that may escape loss of infectivity, since some such residual infectivity is often found (GOSTLING and BEDSON 1956). The question of eclipse needs careful attention. It should be remembered that the evidence for eclipse in phage infection rests both on the true absence of infectious phage and on the genetic evidence that the first infectious phage to reappear is "new" phage.

Experiments where isotopically-labeled influenza virus was used to infect allantoic cells and the isotopes were traced in fractions extracted from the infected cells, have suggested a degradation of virus lipid and a separation of virus constituents, the P^{32} (that is, RNA and lipids) of the infecting virus being found mainly in the nuclear constituents; the S^{35} label (protein) in association with cytoplasmic elements (HOYLE 1957). The probable presence of inactive particles in the inoculum, however, may reduce the significance of these observations (FRANKLIN et al. 1957).

Stronger suggestions for the presence of a noninfectious phase in the multiplication of many animal viruses—and, indirectly for a reorganization of the initial virus material—come from the study of the process of multiplication itself. The evidence as a whole suggests that infectious virus appears by a process of maturation, which is the culmination of a sequence of synthetic processes involving noninfectious vegetative forms.

When influenza and related viruses infect the allantoic cells under conditions of single infection (there are about 10^8 cells lining the allantoic chamber of an 11-day chick embryo), and the allantoic fluid and the allantoic membrane are sampled at intervals and tested separately for virus and virus-related materials, the following observations are made (HENLE 1953). Most virus of the inoculum (70–90%) disappears from the fluid in 20–30 minutes. During the next 2–3 hours, only 1–10% of the adsorbed infectivity can be recovered by extracting the cells. Then, there appear and accumulate in the cells virus-specific substances: S antigen and HA elements with V antigen specificity. These early products are noninfectious; they increase in amounts for several hours and continue to be found even when infectious virus is already present.

An important observation is that the multiplication of influenza virus in the allantoic cells, either *in ovo* or *in vitro,* can be suppressed by enzymatically active ribonuclease provided the enzyme is added to the medium within one hour after infection (LECLERC 1956). Free virus is not affected by ribonuclease. This indicates that in the early stages of infection the virus is modified to a state where its activity can be suppressed by RNase. This state is probably intracellular, insofar as RNase can affect RNA metabolism of intact cells. Later, the viral material reaches a state of integration with the host cell where it is no more accessible or susceptible to the enzyme, or possibly, where the genetic information needed for virus growth is embodied in a non-RNA material.

Infectious influenza virus appears about 4 to 8 hours after infection, depending on the strain used and on the multiplicity of infection, and is liberated for several hours. The yield is only 100–200 infectious units per cell; S antigen may, however, be produced in large excess.

The remarkable thing is that virus particles seem to be formed only at the cell surface, as shown both by titration and by cytological studies. In the first place, practically all influenza virus present in the cells, even one or two hours before the spontaneous virus release, can be freed almost immediately by treatment with RDE (CAIRNS and MASON 1953). Likewise, with NDV growing in cell suspensions, almost all the infectious virus that is still in the cells can be neutralized by treatment with anti-viral serum before extraction (Rubin et al. 1957). Also, release of influenza virus is prevented by the same inhibitor, AMPS, which, as we have already mentioned, prevents virus penetration from proceeding beyond the RDE-sensitive stage (Ackermann and Francis 1954).

In the second place, electron micrographs of cells infected with MNI viruses show no virus particles inside the cells, although S antigen can be demonstrated early in the nucleus by means of fluorescent antibody and V antigen appears in the perinuclear cytoplasm and probably also in the nucleolus (Flewett and Challice 1951, Coons 1957). *Virus particles are seen exclusively at the cell surface,* within a narrow peripheral layer or protruding out of the cell (Morgan et al. 1956).

Thus, the growth of MNI viruses appears to consist of the production of noninfectious virus materials in various parts of the cell, followed by their assembly at the cell surface prior to release. The mechanism of release itself is not clear, but the following facts suggest that surface alteration play an important role. Normal, uninfected allantoic cells emit appendages, which can come loose and be shed into the allantoic fluid (see Burnet 1956). (This material may be responsible for the substrate for virus enzyme that is found free in the allantoic fluid.) In the infected cell this process is exaggerated; the cells show some surface areas lined with complete virus particles, which when sectioned reveal a granular content. Other areas show clumps of long filaments, whose surface resembles that of virus particles, but which are thinner and mostly empty (Morgan et al. 1956). These filaments, once released, have high HA activity.

We get the impression that virus maturation and release represent a profound alteration of normal processes at the cell surface. A reasonable hypothesis seems to be that HA elements, joining with components of the cell surface, form a readily extruded material within which the S antigen, which may be the true genetic material of the virus (Hoyle 1948, Burnet 1956), becomes enclosed. Filaments may be found preferentially in those regions of the cell or with those virus strains where less S antigen is available.

This picture accounts readly for the presence of host-specific antigens in virus particles and filaments and for the low infectivity of the filaments. It agrees with the occurrence in the virus of lipid and polysaccharides whose composition is not very dissimilar from that of similar constituents of cell membranes (Ada and Gottschalk 1956). It raises the interesting possibility that the bond between HA elements and modified cytoplasmic components may be related to that between the virus enzyme and its substrate.

Other interesting observations add to this picture. The RNA composition of the S antigen, extracted from virus, is not absolutely constant; the same virus grown in different host cells contains RNA with different base ratios (Ada 1957). Also, the RNA of free S antigen differs in composition from the one found in the virus. These observations suggest that infection with MNI viruses may give rise to a spectrum of more or less abnormal nucleoproteins. Virus multiplication might be viewed as an RNA pathology in which, under the direction of specific RNA determinants introduced by the virus, new types of nucleoproteins are formed. These would then preside over HA formation and ultimately become assembled together (probably not at random) into mature particles at the cell surface, using as "glue" some more or less modified constituents of the surface of the host cell.

This picture, which is admittedly based only on very circumstantial evidence, fits other observations, including those on incomplete particles and on genetic interactions. The latter will be discussed in a later section.

Incomplete particles (see von Magnus 1954, Burnet 1955) are formed as the main or only product of infection either in cells that cannot support regular virus growth—as with influenza in mouse brain—or, with some strains of influenza, under conditions of heavy infection with large inocula (the von Magnus phenomenon). The incomplete particles are lighter, more collapsible, and contain less S antigen and less RNA per hemagglutinin unit than the infectious particles (von Magnus 1954). When a preparation containing mainly incomplete particles is inoculated in heavy amount into the allantoic chamber, a cycle of incomplete growth follows, giving again a crop of incomplete particles.

It has been suggested that the incomplete particles originate in cells whose surface has been so damaged by multiple infection or other treatments that the assembly process is deranged. Virus particles may exhibit various degrees of incompleteness, which may reflect the presence of different amounts of S antigen per particle (Ada 1957). It is possible that genetic elements, probably S antigenic units, contributed by several incomplete particles may cooperate in initiating infection, giving rise to either complete or again incomplete viruses. Strain differences in tendency to incompleteness may reflect the specific degree of susceptibility of the assembly process to surface damage.

Interference with production of infectious influenza virus by an excess of another MNI virus given during the eclipse period (see Henle 1953) may also be due to prevention of maturation by surface derangement.

In cells of the chorionic layer of the chick embryo one observes with influenza another type of abortive infection, in which apparently the S antigen is the main product, as though maturation failed because of lack of HA synthesis (Fulton and Isaacs 1953, Isaacs and Fulton 1953).

Viruses other than the MNI group.[2] For most other animal viruses, our knowledge of the intracellular growth processes is limited to measurements of infectious virus and to morphological observations, using cytochemical tests, fluorescent antibody, and electron microscopy. Evidence for eclipse,

[2] A report has appeared of the isolation of an infectious RNA fraction from cells of Ehrlich ascites tumor infected (intraperitoneally in mice) with Mengo encephalitis virus. The RNA fraction, inoculated intracerebrally into mice, produced encephalitis, although with an efficiency 1 : 1000 that of the starting virus preparation (Colter et al. 1957). In analogy with findings on tobacco mosaic virus (Gierer and Schramm 1956), to be discussed in a later section, these observations seem to point to RNA as the essential infectious material in these viruses.

as already mentioned, has been obtained with a number of viruses. New intracellular infectious virus appears after various times depending on the virus. Noninfectious, virus-related materials (or "specific soluble substances") are produced beside the regular virus particles in a number of virus infections; but only for the MNI group has their production been correlated with the cycle of virus growth. The virus of WEE in chick embryo fibroblasts is found to rise at about the same time within the cells and in the culture medium, indicating an almost immediate release of newly made virus and, presumably, a surface site of manufacture (RUBIN et al. 1955).

When virus infected cells are studied cytologically, a variety of sequences are observed. Herpes virus antigen is seen first in the nucleus, coincidental with accumulation of Feulgen-positive, DNA-containing material (LEBRUN 1956, SCOTT et al. 1953). Infectious virus, as judged from cell fractionation, is formed in the nucleus and then passes into the cytoplasm, leaving in the nucleus a Feulgen-negative scar, the typical acidophylic intranuclear inclusion. With the electron microscope, new particles are visible first in the nucleus as oval bodies with a single membrane. Later the particles pass into the cytoplasm and finally emerge from the cell as double-coated elements (MORGAN et al. 1954 a). The nucleic acid content of the infectious virus has not been determined. The sensitivity to ether suggests the presence of a lipid component.

With vaccinia virus, and other members of the virus group characterized by particles appearing as brick shaped in dry preparations, the first sign of virus development within the cells of the CAM is often a homogenous or granular matrix, rich in DNA, located near but outside the nuclear membrane (GAYLORD and MELNICK 1953). Within this matrix, there appear later innumerable particles, which seem to go through a process of stage-wise maturation to give the complete, infectious particles. The infectious particles may be liberated in groups or by a simple extrusion process. The virus particles when first seen in the cytoplasm are spheres, about 230 mμ, with a single membrane and an oval "nucleoid." In leaving the cells they seem to acquire a second membrane. The brick shape of free particles may be an artifact, due to the collapse of the spheres over the more rigid oval nucleoid (MORGAN et al. 1954 b).

With viruses of the psittacosis group there has been reported (SIGEL et al. 1951) a process suggestive of binary fission of immature particles. This need not mean that these viruses multiply in a way different from other viruses. It may simply be that the reduplicating elements, instead of being in a morphologically unidentifiable state, are recognizable as units: whether in the reduplicating stage they are infectious or not remains to be decided.

For other viruses, the intracellular location of virus particles has been ascertained. The APC viruses, for example, develop mainly in the cell nuclei, where they produce beautiful crystals of virus particles (MORGAN et al. 1956). Poliovirus is found mainly in the cytoplasm. When the uptake of P^{32} and the synthesis of nucleic acid are followed in various fractions from HeLa cells during infection with poliovirus (type 1), there is observed an early small increase in synthesis of nuclear RNA and DNA, followed by a rapid increase in cytoplasmic RNA synthesis. The rise in this RNA fraction precedes by 1 or 2 hours the formation of intracellular infectious virus, which appears after 3 to 4 hours and is not released until about 6 hours after infection (Loh et al. 1957). The early cytoplasmic RNA presumably is virus-precursor RNA.

A recent report (Baker et al. 1957) has traced the presence of abnormal accumulations of particles, about 15–25 mμ in size, in the nucleolus and cytoplasm of cells infected with measles virus. The presence of these materials is correlated with the appearance of viral antigens. Since the mature virus particles are larger, the intracellular elements may be immature precursors.

It is clear that the process of multiplication of mammalian viruses includes a variety of mechanisms. Yet, in all cases, virus reproduction appears to partake of the quality of synthesis and assembly of cellular constituents rather than of the multiplication of self-contained extraneous parasites. Infectious virus particles arise as the end-product of a series of synthetic processes, in which various constituents are assembled together to give a stable, transmissible element.

It is difficult to ascertain the role of those virus-specific materials that are produced in excess in most virus infections. Whether they are products of virus destruction, defective "rejects" of virus synthesis, or unused precursors is not known with certainly in any one case. Support for the second hypothesis may be found in the already-mentioned differences reported between the RNA of S antigen and of complete virus of influenza grown in allantoic cells (Ada 1957). Definite evidence for or against the precursor hypothesis, which we have tentatively accepted for the S antigen in our discussion of multiplication of MNI viruses, must await detailed studies with isotopic tracers.

The release of viruses other than the MNI group is also a gradual process. Sometimes, the release period continues for hours; with polio virus it is sometimes completed in 2–3 hours (Dulbecco and Vogt 1955). With single polio-infected cells observed in microdrops, release of virus from each cell lasts 1 to 2 hours, is accompanied by vacuolization of a peripheral layer of the cells, and finally leaves a shrunken, scalloped cell corpse (Lwoff et al. 1955).

The above picture of virus multiplication is incomplete in two major respects. First, we lack information as to the form in which the individuality of the particles, their genetic information, is preserved and transmitted through the process of multiplication. The observations on genetic recombination (see below) reveal the existence of discrete genetic properties and thereby suggest discrete determinants, but do not yet throw light on the process of their multiplication. Radiobiological analysis (see Dulbecco 1957) may be of value in estimating the number of "multiplying centers" within infected cells.

Second, we know very little of the specific biochemical processes involved in biosynthesis of animal viruses. Nutritional studies on tissue cultures have revealed only that virus can grow in cells kept in extremely simple media (salts, glucose and glutamine, for example), quite inadequate for cell multiplication (EAGLE and HABEL 1956, RAPPAPORT 1956). Intracellular reserves and probably also products of breakdown of cell constituents must supply materials for viral synthesis.

Inhibitors that block energy production or utilization, such as cyanide, 2,4-dinitrophenol and others, as well as analogues that interfere with the incorporation of amino acids or of nucleotide constituents, generally suppress virus growth (ACKERMANN and FRANCIS 1952; TAMM 1955, HURST and HULL 1956). Methoxinine, for example, can suppress influenza virus even if added several hours after infection. There is little indication of any specificity of requirements among alternative pathways of energy-yielding metabolism between uninfected and infected host cells. Inhibitors of the Krebs cycle enzymes are highly inhibitory for influenza growth in allantoic cells. On the other hand, the growth of feline pneumonitis virus, a member of the psittacosis group, in the yolk sac of chick embryo is not inhibited by malonate (MOULDER 1954). Whether these differences reflect different mechanisms of viral synthesis or differences in host-cell metabolism is not known.

The general indication is that new syntheses of protein and of nucleic acid from nonspecific, low molecular weight constituents are needed for virus formation. These may derive in part from breakdown of host cell constituents, but there is no evidence for a transformation of preexistent host proteins or nucleic acid directly into virus-specific materials. This does not exclude that host-specific components may be incorporated into virus particles, either in the process of assembly, as suggested for influenza, or by incorporation within the virus genome of genetic elements from the host (which might be "copied," however, rather than physically incorporated). We have no genetic evidence for or against such transduction like phenomena. Only the reported dependence of the RNA composition of purified influenza virus on the host tissue (ADA 1957) suggests such possibility.

Fate of the host cells in virus multiplication

The disfunction and death of infected, virus-producing cells are responsible for the characteristic symptoms of viral diseases. Tissues infected with viruses undergo a variety of metabolic alterations, such as reduced or increased glycolysis or changes in certain enzymes (see BAUER 1953). Whatever specificity these alterations may show reflects probably the type of cells affected, the rate of growth and spread of the virus and other secondary processes, rather than any specificity of the primary metabolic alteration in the infected cells. Increases in DNA and RNA synthesis in tissues infected with a virus of the psittacosis group (MOULDER 1954) may be due to secondary cell proliferation, which is observed with many viruses, rather than to a requirement for virus synthesis.

Hence, the knowledge of overall pathogenicity has revealed little about virus growth itself. Pathological changes reflect mainly the virus susceptibility or resistance of various cells and organs, and the extent of virus multiplication in them, which is probably the main factor in pathogenicity. Damage with little or no multiplication of virus may occur under certain conditions, either when excessive amounts of virus may destroy cell mechanisms needed for virus production, or when the virus is "unadapted" to multiply in certain cells, which it can, however, invade and damage (HENLE et al. 1955). Work in tissue culture can clarify some of the mechanisms involved, by defining cytopathogenicity in an operational way and by correlating it with various stages of virus-cell interaction. Thus, damage to a HeLa cell by a single UV-inactivated particle of NDV is proved by the failure of the individual cells to multiply and to form colonies (MARCUS, personal communication). Certain inhibitors, such as *p*-fluorophenylalanine (ACKERMANN et al. 1954) can arrest multiplication of polio virus without preventing cell damage; the cell death may result from presynthetic steps of virus-host cell interaction.

Certain mutants of poliovirus (d mutants; VOGT et al. 1957), exhibit the interesting combination of nonpathogenicity for the central nervous system of monkeys and inability to multiply and cause lesions on monolayers of kidney cells at pH below optimum. The dependence of growth ability and cytopathogenicity on both a genetic and an environmental factor provides an interesting model for the mechanisms that may control variation, tissue specificity and latency in viral infections.

Multiple and mixed infection

The study of the interactions among several virus particles growing in the same cells has revealed complex phenomena of interference, mutual help and genetic cooperation. Infection of animals or of cultured tissues with two unrelated viruses leads sometimes to dual infections, even within the same cell, sometimes to interference, one virus being suppressed by the other (see HENLE 1950). The cellular mechanisms of interference have only begun to be elucidated by work with simple host cell systems. With influenza viruses in the allantoic cells, active or inactive virus may produce interference. As already mentioned, interference may be due to damage to the surface layers of cells, altering the attachment, penetration, and probably also the assembly and release of new virus. The suppression of influenza virus by inactive homologous virus added during the latent period suggests an effect on the latest stages of virus production, possibly on the final assembly.

Clear evidence for mutual exclusion in the early stages is observed with NDV in chick embryo lung cells (BALUDA 1957). In some cases, however, exclusion may occur late and may be specific in its direction. For example, a heat-sensitive polio strain suppresses in mixed infection a heat-resistant mutant, which fails to be produced by most mixed-infected cells. This one-way competition may be due to differential affinity for some special synthetic centers ("key enzymes"?) in the host cell (DULBECCO, personal communication).

When two related viruses succeed in multiplying in the same cell, there are revealed alterations within the virus particles themselves (see Burnet 1955, Hirst 1953, Hirst et al. 1957). One such interaction is "anomalous neutralization," observed in mixed infection of the allantoic cells with two serologically different strains of influenza (or with NDV and influenza A, Granoff and Hirst 1954). Many of the progeny particles are neutralized by serum against either parent strain. Or else, they give particles that can be neutralized by one parental serum but which after growth give rise to progeny of the other parental type. Similar findings have been reported with polioviruses (Sprunt et al. 1955). These phenomena resemble *phenotypic mixing* as observed with phage. They suggest that here too, as in phage, the assembly process can bring together materials synthesized within one cell under the genetic control of several genetically different viral elements. Such process is compatible both with our description of the assembly process of MNI virus and with the idea that small viruses like polio contain a number of protein subunits (Crick and Watson 1957).

In addition to phenotypic mixing, genetic interactions between viruses can lead to a whole series of permanent hereditary virus changes, which can only be attributed to true *genetic recombination*. Such changes have been studied mainly with influenza viruses in a variety of host cells (see Burnet 1955), although analogous findings have been reported for herpes virus (Wildy 1955).

Strains of influenza which differ in a number of properties (antigenic specificity, virulence for various tissues, properties of the viral enzyme, absorbability by aluminum phosphate, etc.) give rise in mixed infection to progeny with novel combinations of properties. Some of the new combinations are unstable; the unstable particles, even when isolated by growth at limited dilution, where only one infectious particle initiates multiplication, give rise to mixtures of parental types. In other cases, however, fully stable hybrids are obtained. Sometimes only one recombinant type is found; but in some instances two reciprocal recombinants are obtained, and, in addition, the original combinations can again be recovered as F 2 recombinants from mixed infection with the F 1 recombinants (Lind and Burnet 1953). This provides evidence for the preservation of discrete determinants throughout these events. Occasionally, "heterozygous" virus particles are observed, which in single infection segregate out two progeny types, usually a parental and a stable recombinant (Hirst 1953).

The characters of influenza A that have been studied in mixed infection experiments behave as if they were determined by two separable groups of genes, each group transferred in block with little or no "crossing-over." Virulence for various tissues, however, behaves as a multi-step character, with a number of variable "dosage" levels (Burnet 1955).

The interpretation of these recombination phenomena in influenza viruses is still obscure. Burnet, who has contributed most of the facts and theories in this field, entertains two main working hypotheses (1955). The first hypothesis assumes haploid particles with two linkage groups plus a variable number of "virulence genes," which reproduce more or less independently of the rest of the genome and can become associated with it in variable numbers. The second hypothesis postulates a polyploid particle, including a number of gene threads or "genomes." In mixed infection, the progeny particles would contain a mosaic of gene threads derived from the two parents. The virulence levels would be determined by the relative numbers of certain gene threads. Preservation of parental types, difficult to visualize in such a situation, would require a tendency for most particles with mixed gene threads to be incomplete and noninfectious. There are, obviously, serious difficulties for both hypotheses, and both are as yet based on insufficient factual material. Quantitative work on the frequency of recombinants, as with phage, will be necessary before a precise theory can be formulated. Also, recombination should be studied using genetically marked strains of known mutant origin. It is possible, although by no means certain, that all recombination phenomena in influenza viruses may be amenable to explanations within the framework of classical genetics.

At any rate, the recombination studies on MNI viruses, coupled with analysis of the RNA composition of various virus strains and mutants and of incomplete virus particles, should ultimately provide an understanding of the form in which the genetic continuity of these viruses is maintained and organized during replication. Already, the findings reveal the complexity of genetic information that can be carried in RNA viruses, a complexity also suggested by the multiplicity of mutations and reverse mutations observed in viruses such as polio, NDV and others. Mutational patterns suggest that the genetic material, at least in poliovirus, is haploid rather than polyploid (Dulbecco 1957).

A detailed genetics of animal viruses, such as we may hope for in the next few years, may well be a major advance in biology, opening up the study of RNA genetics. Other phenomena in virus genetics, such as the transformation of rabbit fibroma virus into myxoma virus in the course of multiplication in the presence of dead myxoma virus or of an extract of it (Berry and Dedrick 1936) may be based on interactions between viral DNA's. It will be important to approach this viral genetics with what might be termed an informed open-mindedness, using known facts of classical and phage genetics as a guide but not as a straight-jacket, and being aware of the possible existence of totally new genetic mechanisms, including interactions between genetic elements introduced by viruses and others already present in the host cells, transfers of information from RNA to DNA and back. It may well be indeed that viral genetics holds the key to the study of gene action and of cell differentiation.

Persistent virus host association in vertebrate cells

Probably the most interesting biologically, although the least understood forms of viral infection, are those that lead to persistent co-existence of virus and host cells. Although the instances are numerous, very few have been studied at the cellular level in a way that can clarify the mechanism of viral persistence, by deciding whether we are dealing with special virus forms such as proviruses, or with infectious viruses with nondestructive properties, or with steady-state balances involving destruction of some susceptible cells, proliferation of refractory cells, and transitions between sensitive and refractory cell states. Information should grow rapidly with the use of the newer methods for the cultivation of single cells and for the measurement of small amounts of viruses. These methods should also help in solving the long debated problems of an acquired cellular resistance to viruses (Enders 1949).

Persistent infections are observed in three groups of situations. First, they occur in intact animals that either have survived acute infections or are symptomless carriers of viruses, whose time and mode of entry may or may not be known. Second, they have been observed in the progeny of cells surviving destructive infection in tissue cultures. Third, they are present in diseases where the cellular response consists of indefinite proliferation with continued virus presence, giving rise to the typical virus tumors.

The state of the virus in natural latent infections has remained unknown, due to the complexity of the systems. Roles of proviruses, of cellular immunity, of balances between thermal inactivation, antiserum action, and viral proliferation have been suggested but not clarified. Some viruses can be carried in asymptomatic condition throughout the life of an animal and can even be transmitted from generation to generation in prenatal life, as with lymphocytic choriomeningitis (Traub 1939); but the mechanisms of virus persistence remain obscure. Persistent infections in simpler systems are a more promising field. Unfortunately, viruses that produce lasting latent infections in animals, such as herpes or choriomeningitis viruses, which ought to be the best candidates for such work, have not yet been investigated for persistent infections in cells cultivated *in vitro.*

Persistent infection in tissue-cultured cells. Several instances have been reported, which illustrate the various possible relations. With the virus of Eastern equine encephalomyelitis (Chalmers 1957), certain pure lines of tumoral cells, following inoculation, undergo waves of partial cell destruction and secondary proliferation, with production of infectious virus in amounts smaller than observed in completely destroyed cell strains. The proliferating cells are susceptible to heavy reinfection from outside, but again give a high degree of survival. Little or no additional virus is revealed by cell breakage. An adequate, but by no means unique, explanation would be the presence of a proportion of virus-susceptible cells produced as physiological variants by a majority of virus-refractory cells. The extent and continuity of virus production would depend on the balance between level of sensitivity of various cells, proportion of cells in different sensitivity states, amount of virus produced and rate of virus inactivation. Rather wide fluctuations would be expected in such a system, and are indeed observed.

With poliovirus, a line of HeLa cells has accidentally been obtained, which multiplies normally only if the medium contains human serum with antibodies against poliovirus (ACKERMANN and KURTZ 1955). When these are removed, the cells undergo rather typical polio-like alterations and liberate some poliovirus type III. The virus from such a culture can be used to re-establish a similar condition. No intracellular infectious virus can be demonstrated in the antibody-grown cultures. The attractive hypothesis that most cells carry the virus in noninfectious form, liberate it in small amounts, and are susceptible to reinfection with larger amounts of it, although by no means the most probable explanation, could easily be tested by the isolation of single cell sublines.

Such isolation has been used in the study of the third example to be discussed, that of NDV in the S-3 line of HeLa cells (CIECIURA et al. 1957). Rare cell survivors from an NDV infection can be cultivated and repeatedly reisolated as single colonies. They are resistant to reinfection with 2 to 4 infectious units per cell, but are susceptible to higher virus doses. They release no detectable free virus in the culture fluid. If, however, they are plated on a layer of X-rayed HeLa cells—which develop into giant cells highly susceptible to NDV infection—each resistant cell acts as a center of infection. The virus thus revealed is apparently normal NDV. The carried condition has not yet been defined nor are data available on the presence of virus in disrupted cells. The peculiarities of liberation of viruses of the MNI group suggest that the blocked step may be virus assembly or release, and that transmission of partly extruded or unfinished virus particles to the hypersusceptible X-rayed cells may occur only by intimate cell contact.

Proliferative cell response and tumor viruses. Proliferation of infected cells is the common response to a variety of viruses, which consequently act as tumor agents. These are found in mammals, birds, and other vertebrates. There is a vast literature on the essential role of the viruses in cell proliferation in the tumors and on the possible role of as yet undiscovered transmissible agents in tumor production (ROUS 1946, OBERLING 1952, BEARD et al. 1955). It is known that some tumor viruses, such as Rous sarcoma virus, can produce destructive responses rather than tumors in hosts of certain ages and genetic background (DURAN-REYNALS 1940). Yet, little work has been done concerning the multiplication of these viruses in the host cells, mainly because of technical difficulties of virus titration. When virus is present in enormous amounts in the tumors, as with rabbit papilloma, its chemical and physical properties can be studied despite the inefficiency of the biological assay system.

A precise assay for Rous sarcoma virus on the CAM of chick embryos has been utilized for a detailed study of the liberation of this virus by established tumor cells cultivated *in vitro* (RUBIN 1955), although not yet for the study of the establishment of infection. The tumor cells liberate virus at an average rate of 1 infectious unit for every 100 cells per hour, slightly more or less depending on the cultural conditions. A stable level of free virus is reached by an equilibrium between virus liberation and thermal inactivation. Breaking of cells does not release free virus. That liberation can occur from practically all cells was shown by experiments with individual cells; release is not by bursts, but by a seemingly random liberation of single infectious particles. The state of the virus in these cells remains unknown.

In summary, persistent virus infections have revealed situations where virus may be maintained by a balance between virus, susceptible cells and refractory or semi-protected cells, and also instances of intrinsic, persistent cell infection compatible with, and even provocative of, cell proliferation. In the cases of NDV and of Rous sarcoma virus, the release mechanisms are apparently quite different. Formally, the NDV system shows how a virus can be carried in uninfectious, inapparent form until exceptionally susceptible host cells become available. The Rous sarcoma system shows that the limited infectivity of a virus-produced tumor may reflect a low rate of virus maturation and liberation, and justifies the hypothesis that potentially transmissible agents in other tumors, if present, may be hard to detect; also, that defective, nonmaturing, but virus-related elements may be present in other tumors. The nature and possible origin of such agents will be dealt with in the closing section.

Multiplication of Viruses in Insects

Insects are known to support multiplication of four groups of viral agents; viruses that cause insect diseases; viruses that are recognized as pathogenic for vertebrates, but can multiply in their insect vectors; plant viruses that multiply in insect vectors; and the agent producing CO_2 sensivity in *Drosophila.* These groups will be taken up in the above order.

Insect-pathogenic viruses

These include a variety of viruses, pathogenic for lepidoptera, hymenoptera and diptera (Bergold 1953; Smith 1955). The best known ones, including the silkworm polyhedrosis virus, consist of rod-shaped or spherical particles containing only DNA and protein and surrounded by proteinaceous membranes (Bergold and Wellington 1956). They are accompanied in the cells of infected larvae by abnormal alkali-soluble proteins, which precipitate either in paracrystalline inclusions or "polyhedra," or as "capsules" around the virus particles. The polyhedral protein, the membrane protein, and the protein of the virus particles are chemically distinct. The absence of precise and sensitive assay methods has precluded quantitative work on infection and viral multiplication, and growth studies are based on electron microscopy almost unaided by other tools. Despite the remarkable iconographic excellence of this work (Bergold 1953, Morgan et al. 1955), one feels somewhat reluctant to accept descriptions of virus growth and development based exclusively on morphological observations. For example, for the virus of silkworm polyhedrosis, which gives intranuclear polyhedra, there is described a series of developmental stages reconstructed from examination of particles trapped in the paracrystalline polyhedra. Minute spheres would evolve into immature rods, which then become mature virus rods (BERGOLD 1953). Virus particles located in the nucleus are often associated with chromatin strands. This association has been interpreted as an essential part of virus multiplication (BIRD 1957), but according to others the chromatic masses in which virus develops arise *de novo* and do not represent altered nuclear chromatin (XEROS 1956). Viruses that give intracytoplasmic polyhedra usually have spherical rather than rod-shaped particles.

Viral latency appears to be the rule in many polyhedral insect diseases; a provirus form, by analogy with prophage, has been postulated in order to explain the wide distribution of symptomless, noninfectious, virus-carrying insect races (VAGO 1951). Induction of provirus in a carrier race may be the cause of the mass virus production and of the disease that follows exposure of some insect races to radiation, peroxydes, nitrite and hydroxylamine (YAMAFUJI 1952). This work, as well as that on virus growth following injection of healthy larvae, awaits adequate quantitation.

A most remarkable virus of the crane fly has been described which is produced in enormous amounts, up to 25% of the weight of the diseased larva (WILLIAMS and SMITH 1957). This DNA virus crystallizes readily and consists of particles which are polyhedral rather than spherical in shape. If a suitable titration method is developed, this virus may prove a choice material for the study of insect virus growth, since its production is not complicated by concomitant synthesis of polyhedral protein.

Viruses of vertebrates and of plants in their insect vectors

A variety of arthropods serve as vectors of viruses pathogenic to animals and plants. The specificity relations reflect a variety of factors, including ecological and biological ones. In a number of instances it has been proved that a virus can multiply extensively in the vector, can be transmitted artificially from one insect to another (or even from one generation to the next) and can grow in cultures of insect tissues. Such observations have been made, on the one hand, with the viruses of yellow fever and equine encephalomyelitis (MERRILL and TEN-BROECK 1935, WITHMAN 1937); on the other hand, with a number of plant viruses (BLACK 1953). A virus can produce the same viral particles and also noninfectious, antigenically specific by-products both in a plant and in an insect vector (BLACK and BRAKKE 1954).

Yet, no studies of the process of multiplication in the insect cells are available and the obstacles to quantitation of these systems appear rather formidable, at least for plant-pathogenic viruses, until some efficient method for viral titration is devised. Especially intriguing is the apparent absence of cytopathogenicity of these viruses for insects, which suggests the possibility of a natural relation between virus and arthropod host.

The Co_2 sensitivity agent of *Drosophila*

A remarkable series of studies has been devoted to a transmissible agent, which controls CO_2 sensitivity in certain races of *Drosophila* (L'Héritier 1951 and 1958). The sensitive flies are paralyzed and killed by an exposure to CO_2 that produces only a reversible narcosis in resistant flies. The agent, called *sigma*, is similar to the more familiar viruses in transmissibility, genetic continuity, dimensions and host specificity. Its behavior, resembling closely that of bacteriophages, reveals a wealth of viral phenomena worth taking as models in guiding work with other viruses. The main facts can be summarized as follows.

The CO_2 sensitive flies contain virus which can be extracted and which, if injected into resistant flies, renders them susceptible. CO_2 sensitivity appears after a latent period of several days (depending on the size of the inoculum), during which the infectious virus (determined by titration of extracts) at first decreases rapidly, then rises to a constant maximum, which is reached at the time CO_2 sensitivity appears. The infectious particles have dimensions of the order of 100–200 $m\mu$; their shape is unknown. Sensitivity is transmitted from a sensitive female to its progeny in one of two ways. A recently infected fly transmits it irregularly, by infection of oocytes as they develop in the ovary; some of the progeny become "stabilized sensitives." If females, the stabilized sensitives transmit the sensitivity in a regular, persistent form to all their progeny; if males, the stabilized sensitives, when mated to a resistant female, give a variable proportion of sensitive progeny, which in turn behave like recently infected flies. In stabilized sensitives, the level of infectious *sigma* is significantly lower than in flies made sensitive by recent infection.

Virus *sigma* is mutable. Among its mutants are found some "defectives," which produce sensitivity but do not give rise to infectious virus (except by another reverse mutation). Host genes controlling receptivity and refractoriness to *sigma* are known, as well as host range mutants of *sigma* able to overcome host refractoriness.

These and other facts fit best the following interpretation. Upon infection of resistant receptive flies, virus *sigma*, after an apparent eclipse, multiplies vegetatively and matures to a maximum. CO_2 sensitivity appears when the amount of vegetative virus has reached a threshold value. The stabilized condition represents a more intimate, persistent relation between the cells and a noninfectious virus form, which is established in females and occasionally gives rise to mature virus. The stabilized form can be transmitted from a female to all offspring, female or male; when, however, it is carried by a male gamete into a zygote it is changed into the unstabilized, vegetative form.

By analogy with phage, the stabilized form would be the provirus, returning to vegetative form and maturation by induction, either spontaneous or upon zygote formation. CO_2 sensitivity would resemble a phage-controlled cell character, present both in hosts with provirus and in hosts with vegetative virus. Immunity, defective proviruses and other features strengthen the analogy. The most glaring difference is the apparent lack of chromosomal location of the provirus *sigma*. Since, however, chromosomal location has been verified only for a few prophages, and for some of them it is a multiple-alternative location, it would not be too surprising if the analogy reached even further. It may well be that the various "forms" that a virus can assume in its association with host cells are representative of the basic alternative states that genetic material can assume in various stages of integration within a functional system.

Multiplication of Plant Viruses

Plant viruses present peculiar limitations and peculiar opportunities for the study of virus growth.

On the one hand, the available host systems are among the least favorable, since they consist of intact plants or plant organs; neither cell suspensions nor pure line tissue culture methods have been successfully developed for plant virus work; this precludes studies of single cycles of infection. Most information derives from mechanical infection of leaves, or, better, of leaf fragments that can be maintained floating on nutrient solutions for many days. In addition, titration methods for plant viruses are most inefficient, even for viruses that give local necrotic lesions on inoculated leaves, as tobacco mosaic virus (TMV) does on *Nicotiana glutinosa.* The usual ratios "infectious units/virus particles" are of the order of $1:10^6$ or lower.

Infection can only start when virus enters though the wounded surface of an epidermal cell. The persistence of residual, nonmultiplying inoculum on inoculated leaf tissue precludes precise study of the early part of viral multiplication.

On the other hand, knowledge of the chemistry and physics of some plant viruses has developed rapidly because of the large amounts in which viruses are available, of their stability, their ease of purification, and their crystallizability. All plant viruses that have been purified thus far seem to contain RNA and protein only. The chemical approach has contributed some insight into the role of nucleic acid and protein in initiating infection and into some stages in virus assembly. Knowledge in this field is growing rapidly.

TMV is the favorite experimental material; it is present in enormous amounts in infected tissues, up to a few per cent of the dry weight. A single large infected cell may contain as much as 6×10^7 virus particles, of the order of 100 or more per μ^3 (Nixon 1956). Other viruses that have provided information on virus growth include turnip yellow mosaic, tobacco necrosis and several others less intensively studied.

Little is known yet about the growth mechanisms of viruses that can be transmitted by insect vectors only; some of these viruses, as already mentioned, multiply both in the plant and in the insect host and insect tissues may provide a more favorable host system for the study of their multiplication (see Black 1953, Maramorsch 1955).

The nature of the infectious material and the initiation of infection

Only those properties of virus particles directly relevant to virus multiplication will be mentioned here. Infectious preparations of TMV contain particles of hexagonal cross-section, of variable length and uniform thickness. The amount of infectivity parallels the proportion of particles 300 mμ long, which are probably the only fully infectious ones (see Williams 1957), although this has long been disputed (see Bawden 1957). The particles contain 5.6% RNA (about 2.7×10^6 mol. wt. units), the rest being protein. The protein moiety consists of apparently identical subunits, 17,000 mol. wt.. of which there are about 2000 in a particle 300 mμ long (Knight 1957). The protein subunits are assembled helically around a hollow core (R. E. Franklin et al. 1957). The RNA is located at about half-way the radius of the particles, protected from RNase activity, and possibly at the bottom of an internal groove formed by the protein subunits.

Protein and RNA are readily separated by alkali or detergents (Schramm 1947 a, b). The protein obtained by alkali treatment is in form of small fragments, hollow-centered discs or short rods and has a tendency to reassemble in longer rods of the same thickness as virus particles as the pH is lowered. The protein fraction from virus is not infectious. It resembles the virus-related proteins that are found in infected plants along with the virus (see below).

When alkali-extracted protein and detergent-extracted RNA are mixed, they reaggregate readily (at pH higher than needed for protein reaggregation), and reconstitute particles of variable length. The preparations of reconstituted particles are infectious, although their infectivity per μg. RNA is usually not more than a few per cent of that of the native virus (Fraenkel-Conrat and Williams 1955, Commoner et al. 1956). Finally, RNA preparations are themselves infectious, although with an even lower efficiency (Gierer and Schramm 1956).

These findings emphasize the central role of virus nucleic acid in the initiation of infection. This role is supported by the knowledge that extensive changes in the protein moiety of the virus (such as the removal of all L-threonine end-groups, the blocking of the —SH groups of cysteine, and many others) are compatible with full infectivity (see Knight 1954 and 1957), while chemical changes in the RNA reduce infectivity (see Jeener 1956). Also, virus reconstituted from the protein of the regular TMV strain and the nucleic acid of the ribgrass strain gives rise to ribgrass virus, proving the role of nucleic acid as carrier of genetic information (Fraenkel-Conrat 1956).

Hence, the RNA of TMV is now generally visualized as the *primum movens* of infection; the protein would exert an accessory function, probably a protective and stabilizing one both within the host and outside. Two kinds of objections may be raised, however. First, with regard to the infectivity of reconstituted virus preparation, it has been pointed out (BAWDEN 1957) that the protein fragments are inhibitory of infectivity and could mask the infectivity of a small residual proportion of undegraded virus. Reaggregation could then reveal this residual virus by reducing the inhibitory effect of the fragments. This objection should not apply to the infectivity of the naked, purified nucleic acid.

Second, it might be suggested that since usually only one out of every 10^6 particles can be surely considered as infectious, the RNA from the bulk material may not be representative of the infectious particles, which might even contain something else altogether, e. g., DNA. This question may be resolved either by devising more sensitive infectivity tests or by more extreme purification. In what follows we shall discount these objections (without necessarily forgetting them) and assume that reconstituted virus has infectivity and that the essential nucleic acid is indeed RNA.

Although successful reconstitution studies have been limited to TMV, other viruses give similar evidence for the primacy of RNA in reproduction. Thus, of the two types of spherical particles found in plants infected with turnip yellow mosaic, only the ones with RNA are infectious; the others, similar in size but RNA-free, are not (MARKHAM and SMITH 1949).

The characteristics of the RNA that are required for infectivity are not yet certain. Depolymerization of the naked RNA by RNase destroys, of course, its infectivity, whereas the native virus is generally resistant to the enzymatic action of RNase (which, however, can give some virus inhibition by combining with it; seee BAWDEN 1950). That the whole RNA of an infectious, full length particle may be required is suggested by the fact itself of the greatest (and probably exclusive) infectivity of such particles, and in addition, by X-ray irradiation studies. These give the same value for the X-ray sensitive volume of TMV and of naked RNA from TMV (GINOZA and NORMAN 1957). This sensitive volume corresponds roughly to that of the whole RNA of a particle 300 mμ long. The known tendency of RNA to polymerize might, however, be partly responsible for the apparent size of the infectious RNA elements. The sensitivity of naked RNA to ultraviolet light is also identical to that of intact virus (SIEGEL et al. 1956), except for a curious observation. A TMV strain that is 5.5 times more UV-resistant than the usual strains yields RNA which, naked or reconstituted with protein, is as sensitive as that from sensitive strains. This suggests either a role of the protein in determining the molecular configuration of the RNA or, possibly, a different requirement for structural integrity of this RNA when it enters the host cells naked instead of protected by its protein coat.

Experiments on TMV with incorporated 2-thiouracil (VAN RYSSELBERGE and JEENER 1957; see below) have been interpreted as suggesting that the infectious particle contains about 8 equivalent RNA elements, any one of which is sufficient to initiate infection. The problem of the size of the infectious unit of RNA, and of its relation to the variable infectivity of viral preparations (BAWDEN 1957), requires further study.

Overall course of virus multiplication

The production of TMV in recently infected leaves has been studied by a variety of direct methods, including biological assays of infectivity, electron microscopy, serological assays of viral antigens, and chemical assays of viral protein (COMMONER et al. 1950, STEERE 1952; YARWOOD 1952). The results vary with the method and the environmental conditions. Infectivity measurements reveal virus increases earlier than other methods. Because of their relatively low sensitivity and of the presence of residual virus inoculum, they fail to give definite evidence for or against an early eclipse period. Even the existence of a period before the first rise in infectious virus titer is not well established, but such a period may last several hours. Numerous conditions, such as the age of the host cells, their metabolic status, temperature, light and moisture modify virus production. No illuminating observations have emerged, and the facts as a whole mainly recommend caution in interpreting isolated observations under poorly controlled conditions (BAWDEN and KASSANIS 1949).

Better information on the early stages of infection and on the mechanisms of virus biosynthesis comes from indirect methods.

The early stages of infection. Infection starts with the penetration of infectious material into a susceptible cell. For TMV in leaf cells of tobacco, this requires, apparently, penetration into wounded epidermal cells. Infiltration of leaves with TMV through the stomata, by which virus reaches parenchymal cells, gives no infection, although labelled C and P atoms from the virus inoculum are found associated with insoluble cell constituents (JEENER and VAN RYSSELBERGE 1955). Penetration of virus into the cells may not occur under these conditions, however.

Two different approaches both suggest an early change in virus organization. First, RNase infiltration of tobacco leaves, either previous to mechanical infection with TMV or within 15 minutes after it, completely suppresses virus multiplication (HAMERS-CASTERMAN and JEENER 1957). By two hours, the infection has reached an RNase-resistant condition. RNase apparently acts enzymatically to produce this effect; peroxide-inactivated RNase is ineffective.

Second, the changes in the UV sensitivity of TMV within recently infected leaves of *Nicotiana glutinosa* suggests early alterations in virus structure (SIEGEL and WILDMAN 1956). The UV sensitivity is followed by counts of local lesions, after irradiation with graded UV doses at various times. The survival curves of "infectious centers" serve as indications of the UV susceptibility of the infecting materials or their progeny. After an early period of about two hours, in which the UV sensitivity remains constant and similar to that of free virus (or free viral RNA), the infectious centers become progressively more resistant for about three hours, although the killing curve remains a first order one, suggesting a single-hit mechanism. Then, the resistance increases very rapidly and the curves become of multiple-hit type. At higher temperatures, still in the range where virus grows, these changes take place more rapidly.

How are these observations to be interpreted? Does each infecting virus particle pass through various stages in its integration with the initial host cell, including an RNase-sensitive, naked RNA state? Does virus multiplication take place by replication of such naked, vegetative, possibly noninfectious elements, with maturation reconstituting the protein-coated, RNase-resistant particles? The UV sensitivity cannot reflect only the accumulation of vegetative virus; it must express a change in state, either the penetration of virus into sheltered regions, or the formation of smaller, more resistant units, or else the appearance of UV-resistant centers of virus production, possibly by transfer of the virus-specific genetic information to materials other than nucleic acid.

The need for the transformation of the infecting virus particles into a naked RNA form, which may be the replicating form, is confirmed by similar experiments of UV irradiation of leaves infected with naked RNA from TMV (SIEGEL et al. 1957). Here, the changes in UV sensitivity start with little or no lag and follow the same course that would be observed about one hour later in infection with intact virus. Incidentally, these results provide indisputable evidence, if any were needed, for the difference of the infection-initiating material in the two cases.

Similar, although less precisely investigated stages in the evolution of virus sensitivity to RNase and to UV light have been reported for tobacco necrosis virus in bean leaves (BAWDEN and HARRISON 1955). With potato virus X, which, after being rendered inactive with UV light, can be reactivated within the host plant by visible light (BAWDEN and KLECZKOWSKI 1955) in the same way as phage is (DULBECCO 1950), the pattern of photoreactivability also suggests a series of stages, with a relatively brief period of photoreactivability, after which the irradiated virus inoculum, if not reactivated, is irretrievably lost.

It is important to remember that what is observed in these studies is the final outcome of an infection that spreads from one to many cells, rather than the outcome of infection in a single cell. We do not know whether a vegetative, naked virus form would be capable of migrating as such from cell to cell or whether cell-to-cell spread occurs only in the form of mature, complete particles. Until these and other questions are answered, it will be difficult to decide how much of the stagewise evolution of early infection can be interpreted in terms of cell-virus interaction and how much in terms of virus spread.

Mechanisms of normal virus biosynthesis. As with other groups of viruses, the protein and nucleic acid of TMV appear to be formed almost exclusively from constituents assimilated *de novo* after infection, not by incorporation of preexistent cellular macromolecules (MENEGHINI and DELWICHE 1951, COMMONER and DIETZ 1952). The most readily available external N source is ammonia. The amino acids used in virus protein are apparently not in equilibrium with the free amino acid pool of the tissue (COMMONER et al. 1953).

Once made, the virus particles are stable and little or no turnover of accumulated virus takes place (MENEGHINI and DELWICHE 1951). This does not prove that new mature virus is not responsible for the transfer of infection from cell to cell or that it can initiate infection in new cells without being broken down. The amount of virus involved in the spread would probably be undetectable by the isotopic methods.

Important information on virus biosynthesis comes from studies of virus-related materials, which indicate the presence of noninfectious, transient precursors of macromolecular size. In the first place, virus-infected tissue accumulates for several days RNA in excess over the amount accounted for by the complete virus; this excess RNA finally disappears (BASLER and COMMONER 1956). More definite information derives from the study of easily identifiable, virus-related proteins.

Several abnormal protein fractions are present in TMV-infected tissues. The most important ones, variously referred to as X-protein, B 3-B 8 protein, soluble antigens (COMMONER et al. 1952, TAKAHASHI and ISHII 1953, JEENER and LEMOINE 1953) consist of RNA-free viral protein in various states of aggregation, and with the same ability to aggregate with viral RNA as the protein extracted from virus (JEENER 1956). These noninfectious proteins usually represent only a small proportion of the total virus protein in a heavily infected plant, around 5% at most. Isotope experiments, using short exposures to $C^{14}O_2$, show that the noninfectious proteins become transiently labeled (that is, are synthesized from external C) more rapidly than virus. The rate of appearance of the labeled C in the virus is compatible with a steady-state situation in which the noninfectious protein acts as a necessary intermediate, synthesized first and then incorporated into complete virus (VAN RYSSELBERGE and JEENER 1957). Experiments with N^{15}-labeled ammonia (DELWICHE et al. 1955) give results compatible with this view.

These observations seem to establish that at least part of the noninfectious protein is a precursor of virus particles. What we know of the aggregative tendencies of this protein makes it plausible that it can readily combine with newly formed virus RNA to make complete virus. It is unlikely, however, that all noninfectious protein is precursor protein; more probably, some of it aggregates alone to form stable, noninfectious byproducts.

Two points must be remembered. First, conditions in the cells are certainly different from conditions *in vitro* with regard to aggregation of viral constituents, and may lead to different selectivities of aggregation. Indeed, even the TMV particles within the cells can form, together with other materials, crystals very different in shape and internal organization from the paracrystals of pure virus (STEERE and WILLIAMS 1953). Second, if complete virus is assembled by aggregation of RNA with protein made separately rather than around it, then there may be a certain variability in the composition of the complete virus particles themselves, especially reflecting physiological conditions in the host cells. This may be partly the cause for the variable RNA content and specific infectivity of various TMV preparations (BAWDEN 1957).

With turnip yellow mosaic virus, the uninfectious, RNA-free particles incorporate C^{14} from CO_2 faster than the infectious particles (JEENER 1956). It is difficult to visualize these spherical, probably hollow particles as precursors, however; more likely, they may stem from protein synthesized when RNA is in short supply.

The above indications of an assembly of viral protein with RNA as a terminal phase in the formation of complete particles suggest the possibility of observing with plant viruses both the incorporation of host RNA into virus (as in transduction by phage) and the incorporation of mixtures of two different proteins in mixed infected cells (as in phenotypic mixing). Directly relevant experiments have not yet been reported. A suggestive possibility, however, is indicated by the report (WANG and COMMONER 1956) that TMV protein can be aggregated with DNA extracted from tobacco plants, either normal or TMV infected. DNA of other origin seems not to have been tried. Remarkably, the DNA-protein complex has some infectivity, which seems to arise with the reaggregation. Admittedly, the possible presence of some residual viral RNA, which could be made infectious by incorporation along with DNA into stable particles, has not been excluded. Also, complications due to interference by disaggregated protein with the infectivity of small amounts of RNA, which might thus escape detection (BAWDEN 1957) may affect these as well as other findings on reaggregated infectious virus. Yet, this finding of DNA-protein aggregates requires careful attention. Especially, any possible role of DNA specificity, depending on source of origin and mode of preparation, should carefully be investigated.

Changes in virus RNA composition. 2-Thiouracil is a powerful inhibitor of TMV multiplication in plant leaves (COMMONER and MERCER 1952). Several other purine and pyrimidine analogues are also inhibitory, though generally less so (MATTHEWS and SMITH 1955). The greater effectiveness of thiouracil may relate to an apparent limited availability of uracil for virus biosynthesis (BASLER and COMMONER 1956).

The important fact is that thiouracil—as well as azaguanine and possibly other analogues—is actually incorporated into the RNA of TMV (JEENER 1956). Thiouracil may replace as much as 5 to 18% of the viral uracil. This incorporation is not wholly surprising, because thiouracil and azaguanine would probably fit into any probable macromolecular configuration of RNA, just as 5-bromuracil fits into the structure of DNA, replacing thymine (DUNN and SMITH 1954). Enzymes exist that can synthesize the nucleosides of many purine and pyrimidine analogues (FRIEDKIN 1954).

The analogue-containing TMV is less infectious than the normal virus. This low infectivity must be responsible for the complete suppression of infection by early administration of thiouracil, which can be relieved competitively by uracil if this is given early, but not after a few hours (COMMONER and MERCER 1952). This suggests that by this time the initial infectious virus is no more available and that the new "mislabeled" virus is fully inactive.

The thiouracil-containing virus produced when infection is not completely suppressed has low infectivity, as judged by the amount of new virus it can give rise to within a given time in tobacco leaves, but can give rise to an almost normal number of local lesions on *Nicotiana glutinosa* (Van Rysselberge and Jeener 1957). This has suggested that each virus particle contains several "genomes," one of which at least must be functional to initiate a lesion on *Nicotiana glutinosa*. The amount of virus produced in tobacco, on the other hand, would reflect the total number of functional genomes in the inoculum particles. The number of genomes per particle has been calculated as 8; one genome would be nonfunctional if, on the average, it contained 16 thiouracil-nucleotides. This explanation is clearly not the only possible one. We might conceive, for example, that the virus particles have a single genetic element, which when loaded with thiouracil may still be infectious (as observed in *N. glutinosa*), but can initiate infection only slowly, possibly more so in tobacco than in the other hosts.

Also, the hypothesis of multiple infectious elements seems difficult to reconcile with the X-ray sensitivity data, which give a sensitive volume equal to the total RNA (Ginoza and Norman 1957); but the damage produced on a bundle of RNA molecules by X-rays may be qualitatively different from the damage due to abnormal nucleotides. Yet, the concept of a multiplicity of RNA strands, each potentially infectious, seems in agreement with some other recent observations (Commoner et al. 1955). When RNA from TMV is fractionated by absorption methods and the fractions are reconstituted with virus protein, certain fractions give an infectious material that gives rise to new, abnormal symptoms. This might reflect a selective concentration in these fractions of certain RNA units, which were mixed with other units in the initial virus. Clearly, such a multiplicity of infectious elements would allow for genetic inhomogeneity, and hence for genetic recombination by reassortment of non-identical elements.

Changes in the biological properties of viruses. Mutations in TMV and other plant viruses are apparently common; the composition of virus strains presumably of mutant origin has been investigated in detail (Knight 1954). Biologically, two variation phenomena, besides the one mentioned in the preceding section, may provide clues to viral multiplication.

The first is the discovery of a virus strain, cowpea mosaic, related to TMV, which has different properties, both biological and physical, depending on whether it has multiplied on tobacco or bean plants (Bawden 1956). The changes are fully reversible upon transfer from one plant to another. Although a system of mutation and reverse mutation is not excluded, the observations suggest a host-controlled change in virus properties, remindful of host-induced modification in phage (Luria 1953 a). Further studies may reveal interesting interactions between viral and host RNA.

The second instance is the production of stable strains with new combinations of characters, mainly expressed as types of lesions, in plants infected with two strains of tomato spotted wilt virus differing in several characters (Best and Gallus 1955). The "recombinants" breed true. They are found both in tomato and in *N. glutinosa* plants. They certainly do not represent frequent mutants; but complex selections cannot easily be excluded, especially in work on recombination between strains of unknown origin.

Persistent infections with plant viruses

Virus infected plants can live and grow carrying virus in most cells. Some viruses are excluded from the gametes, but others are more or less regularly transmitted through the egg cell, less frequently through the pollen. Occasionally, plants recover from a virus infection and prove immune to reinfection with the same virus or related viruses (PRICE 1940, BENNETT 1953). This acquired immunity is unexplained; virus is generally present in reduced amounts in the immune plants, including the new tissues if still immune. The suggestion has been made (BERTANI, personal communication) that this condition of acquired immunity may be analogous to lysogeny in bacteria, being controlled by a noninfectious "provirus," which occassionally gives rise to some infectious progeny. Evidence for this view is lacking and will be difficult to obtain until pure lines of cells can be grown from infected plants and tested at the cellular level. Lysogeny in bacteria, we must remember, is strictly a cellular condition and the provirus state of a virus is only defined and verifiable at the cellular level, from evidence of virus production from the progeny of cells that contain no infectious virus. Whether this or other forms of carrier state occur in plants remains to be seen.

The relation of the virus to the cell

"All science is the search of unity in hidden likenesses."
J. BRONOWSKI. *Science and Human Values, The Nation,* December 29, 1956.

The mechanisms of virus multiplication as at least partially understood today reveal a variety of patterns, but by no means a bewildering variety. Regularities appear when virus multiplication is considered at the genetic and molecular levels. What generalizations do these regularities justify? And what do they tell us about the nature of viruses and their place in the world of living things?

In an attempt to answer these questions, we shall review the main conclusions reached in the preceding sections and accept some of them as proven even though their validity may still be open to serious question.

Phage multiplication can be considered as being essentially identical with DNA replication. Whatever steps in phage replication may involve non-DNA elements, they are probably common features of the replication of all DNA. Indeed, phage is providing the most direct approach both to the mechanisms of DNA replication and to the chemical basis of genetic specificity and function.

The phage genome can multiply either in coordination with or to the exclusion of the host cell DNA, that is, either as prophage or as vegetative phage. Not only vegetative phage, but also prophage, and indeed the whole bacterial "chromosome," behave essentially as DNA, at least in their susceptibility to decay of radioactive atoms (Sent et al. 1957). Phage maturation should then be considered as one of several possible expressions of the genetic function of phage DNA, just as any developmental process is an expression of the function of genetic material.

Not enough is known about the multiplication of any animal virus containing DNA to indicate whether or not the viral DNA plays a role similar to phage DNA. RNA viruses, however, are somewhat better known. Here there are reasons to believe that the growth-initiating element consists of pure RNA for some viruses (prototype: TMV) or of a RNA-protein complex for other viruses (prototype: influenza). The mature TMV particle is certainly not a "ribonucleoprotein" in the sense of a specific functional complex of coded RNA and coded protein (if any such complex exists); the S antigen of influenza might, instead, be such a ribonucleoprotein. Whether the RNA of TMV must become part of a ribonucleoprotein in replication, or whether the synthesis of a specific DNA is involved as an intermediate step in the replication of TMV-like and of influenza-like viruses remains to be seen. At any rate, we are inclined to consider the protein of TMV particles and the hemagglutinin and lipid-containing fractions of influenza as developmental products of the function of the virus genome, involved in virus maturation rather than in replication, in the same way as the proteins of phage particles.

According to this interpretation, the process of virus replication is reduced to the biosynthesis of coded nucleic acid macromolecules and virus maturation is identified with the control of cell products through genetic function. The proplem of nucleic acid biosynthesis is becoming amenable to experimental study (see Kornberg 1957). That of genetic control of development, which must involve transfer of specificity between macromolecules, may also become clarified in the near future.

The virus-infected cell replicates new genetic elements and exhibits new genetic functions. Thus, we visualize infection as a *genetic infection*, which may take the form of genetic parasitism or of genetic integration, depending on the functional compatibility of the participating components. It was pointed out some years ago (Luria 1953 b) that the essential feature of virus infection is the *merging* of genetic elements within a cell. We observe as viruses those genetic elements which are capable of transfer from cell to cell and which are prepared for such transfer by a specific process of maturation within the host cell. This process resembles, of course, that of sexual fertilization, in which the whole nucleus of a cell is prepared for introduction into another cell. It also belongs with a host of other phenomena, in which intracellular parasitism can give rise to more or less stable integrated systems (Lederberg 1952).

What genetic elements are those brought into the cell by the infectious virus particles? Are they complete strangers to the recipient cell? Or are they more or less normal elements of cells, which have developed peculiar mechanisms for transfer? And if so, how exceptional are these infectious cell constituents?

The answers seem relatively easy for phage DNA. This is apparently a close relative of the DNA of the bacterial cell. As prophage, it can associate with the bacterial "chromosome." Whether the prophage sits linearly within a completely linear DNA chromosome, or is attached as one or more branches to a cell chromosome (see BERTANI 1958)—which might actually possess many such branches of its own—a prophage seems to be simply a fragment of bacterial DNA which can multiply vegetatively and can build a peculiar and specific apparatus for its own transmission.

How many portions of the bacterial DNA have such transmissibility cannot yet be said. Defective prophages, virulent and intemperate phages, and bacteriocines (see JACOB 1954), may represent a series of genetic variations on the prophage theme. The transmissibility of artificially extracted DNA in transformation phenomena indicates that conditions in bacteria are such as to make the most of any natural mechanism that can protect DNA and help its transfer from cell to cell. This clearly happens in transduction of host properties by phage. Phage maturation and cell mating (WOLLMAN et al. 1956) are thus two ways in which DNA transfer can be accomplished by bacteria in nature. Both result in partial merging of genetic systems and thus belong to a group of phenomena that have been called "meromixis."

Let us now turn to RNA viruses. Here a lead may be provided by comparing the RNA-containing elements of the viruses of the influenza group (the S antigens) with the microsomes of animal cells. The microsomal particles in cell extracts have sizes of the same order as the S antigen. They also contain RNA and protein. They are involved in the synthesis of cellular proteins (ZAMECNIK et al. 1956, SIMKIN and WORK 1957). In the living cell the microsomes are apparently organized in what is called the "endoplasmic reticulum" (PALADE 1956). It is tempting to consider the RNA-protein constituents of the influenza group viruses as infectious microsomes and the maturation of these viruses as the expression of an abnormal pattern of protein synthesis. Other groups of viruses may, likewise, be carriers of microsomes with different patterns of maturation. With viruses such as TMV, where RNA alone seems to control infection, the RNA might possibly function either by merging with host microsomes or by becoming a microsome through the initiation of specific protein synthesis. If this were so, then the RNA would acquire the same primacy as carrier of specificity in the microsome that DNA appears to have in the chromosome.

Such a hypothesis, admittedly based on very thin evidence, would attribute to the microsomal elements a degree of genetic autonomy that cannot be verified easily by the study of cells as a whole (see EPHRUSSI 1953). There is no difficulty here, however, since the very existence of RNA viruses is proof of the intrinsic genetic specificity of RNA. Nor is there any reason to suspect that one category of RNA may be genetically more autonomous than another. Whether RNA is or is not, as has been suggested (STENT 1957), a necessary intermediate in DNA replication, it still remains the leading candidate to fulfill in the cell the role of delegated agent of genetic function, and as such would still retain the essential pattern of specificity. It may indeed be difficult to decide as to whether the primacy in genetic specificity belongs to DNA or RNA; posed in this way, the problem may ultimately prove to be a pseudoproblem.

To homologize, even in a purely hypothetical way, viral DNA with chromosomal DNA and viral RNA with microsomal (or other) RNA, does not mean to assert that all viruses as we know them now arise directly by the acquisition of transmissibility by cell constituents. A great deal of independent evolution may have rendered present day viruses quite different from the analogous constituents of the cells in which they can grow. It seems reasonable to suppose that the greater the genetic similarity, the greater will be the chance for a persistent functional association between a virus and its host cell. The most frequent outcome of infection, however, will be virus reproduction and cell destruction, presumably because of the relatively unspecific range of host cells that the transmission mechanism makes available to a virus.

What would seem to be new viruses might arise by acquisition or reacquisition of transmissibility by cell constituents, which may have derived their specificity either from purely endogenous directives or from previously acquired extraneous elements. Indeed, viruses, when viewed as functional cell constituents, raise—and partially answer—the question of the possible origin of cells by the merging of genetic lines, through infectious heredity, rather than only by the progressive differentiation of genetic elements within a single line of descent. Regressive evolution through parasitism (see Burnet 1955), although unlikely to account directly for the origin of viruses such as phage or the RNA viruses which have provided the basis of our speculations, may have played a role either indirectly, as a mechanism for genetic merger of cell constituents, which may then become viruses, or possibly more directly, in giving origin to some of the viruses whose multiplication is still unexplored.

Yet, whatever mechanisms future discovery may reveal, it seems now a fruitful unifying view to consider viruses as genetically specific cell constituents containing coded DNA or RNA, which can, as one of their genetic functions, determine their own incorporation into specific vehicles for transmission to other cells.

References

Ackermann, W. W., and T. Francis Jr., 1954: Characteristics of Viral Development in Isolated Animal Tissues. Adv. Virus Res. **2**, 81—108.

— and H. Kurtz, 1955: Observations Concerning a Persisting Infection of HeLa Cells with Poliomyelitis Virus. J. exper. Med. (Am.) **102**, 555—565.

— and F. E. Payne, 1957: Action of Immune Serum upon a Cell-Virus Complex. Fed. Proc. **16**, 404.

— A. Rabson, and H. Kurtz, 1954: Growth Characteristics of Poliomyelitis Virus in HeLa Cell Cultures: Lack of Parallelism in Cellular Injury and Virus Increase. J. exper. Med. (Am.) **100**, 437—450.

Ada, G. L., 1957: Ribonucleic Acid in Influenza Virus. In: The Nature of Viruses. Ciba Foundation Symposium, London.

— and A. Gottschalk, 1956: The Component Sugars of the Influenza-Virus Particle. Biochem. J. **62**, 686—689.

ADAMS, M. H., 1954: Abortive Infection with Viruses. In: HARTMAN, HORSFALL, and KIDD. The Dynamics and Rickettsial Infections. New York.
— and B. H. PARK, 1956: An Enzyme Produced by a Phage-Host Cell System. II. The Properties of Polysaccharide Depolymerase. Virology 2, 719—736.
— and F. E. WASSERMANN, 1956: Frequency Distribution of Phage Release in the One-Step Growth Experiment. Virology 2, 96—108.

ANDERSON, S. G., 1954: The Growth Curve of Vaccinia Virus on the Chorioallantois. Austral. J. exper. Biol. a. Med. Sci. 32, 633—639.

ANDREWES, C. H., and D. M. HORSTMANN, 1949: The Susceptibility of Viruses to Ethyl Ether. J. gen. Microbiol. 3, 290—297.

ARBER, W., G. KELLENBERGER et J. WEIGLE, 1957: La défectuosité du phage lambda transducteur. Schweiz. allg. Path. Bakt. 20, 659—665.

AVERY, O. T., C. M. MACLEOD, and M. MCCARTY, 1944: Studies on the Chemical Nature of the Substance Inducing Transformation of Pneumococcal Types. J. exper. Med. (Am.). 79, 137—157.

BAKER, R. F., F. RAPP, E. A. GROGAN, and I. GORDON, 1957: Visualization of Measles Virus in Human Cells. Bact. Proc. 10, 76.

BALUDA, M. A., 1957: Heterologous Interference by Ultraviolet-Inactivated Newcastle Disease Virus. Virology 4, 72—96.

BARNER, H. D., and S. S. COHEN, 1954: The Induction of Thymine Synthesis by T 2 Infection of a Thymine Requiring Mutant of *Escherichia coli*. J. Bact. 68, 80—88.

BASLER, E., Jr., and B. COMMONER, 1956: The Effect of Tobacco Mosaic Virus Biosynthesis on the Nucleic Acid Content of Tobacco Leaf. Virology 2, 13—28.

BAUER, D. J., 1953: Metabolic Aspects of Virus Multiplication. In: The Nature of Virus Multiplication. Symp. Soc. Gen. Microbiol. Cambridge, England.

BAWDEN, F. C., 1950: Plant Viruses and Virus Diseases. Waltham, Mass.
— 1956: Reversible, Host-Induced Changes in a Strain of Tobacco Mosaic Virus. Nature 177, 302—304.
— 1957: The Multiplication of Plant Viruses. In: The Nature of Viruses. Ciba Foundation Symposium, London.
— and B. D. HARRISON, 1955: Studies on the Multiplication of a Tobacco Necrosis Virus in Inoculated Leaves of French-Bean Plants. J. gen. Microbiol. 13, 494—508.
— and B. KASSANIS, 1949: Some Effects of Host-Plant Nutrition on the Multiplication of Viruses. Ann. Appl. Biol. 37, 215—228.
— and A. KLECZKOWSKI, 1955: Studies on the Ability of Light to Counteract the Inactivating Action of Ultraviolet Radiation on Plant Viruses. J. gen. Microbiol. 13, 370—382.

BEARD, J. W., D. G. SHARP, and E. A. ECKERT, 1955: Tumor Viruses. Adv. Virus Res. 3, 149—197.

BENNETT, C. W., 1953: Interactions between Viruses and Virus Strains. Adv. Virus Res. 1, 39—67.

BENZER, S., 1955: Fine Structure of a Genetic Region in Bacteriophage. Proc. Nat. Acad. Sci. 41, 344—354.
— 1957: The Elementary Units of Heredity. In: MCELROY and GLASS, The Chemical Basis of Heredity. Baltimore.

BERGOLD, G. H., 1953: Insect Viruses. Adv. Virus Res. 1, 91—139.
— 1953: The Multiplication of Insect Viruses. In: The Nature of Virus Multiplication. Symp. Soc. Gen. Microbiol., London.
— and E. F. WELLINGTON, 1954: Isolation and Chemical Composition of the Membranes of an Insect Virus and their Relation to the Virus and Polyhedral Bodies. J. Bacter. (Am.) 67, 210—216.

BERRY, G. P., and H. M. DEDRICK, 1936: A Method for Changing the Virus of Rabbit Fibroma (Shope) into that of Infectious Myxomatosis (Sanarelli). J. Bacter. (Am.) 31, 50—51.

BERTANI, G., 1953: Lysogenic versus Lytic Cycle of Phage Multiplication. Cold Spring Harbor Symposia Quant. Biol. 18, 65—70.
— 1954: Studies on Lysogenesis. III. Superinfection of Lysogenic *Shigella dysenteriae* with Temperate Mutants of the Carried Phage. J. Bacter. (Am.) 67, 696—707.
— 1957: A "Dismune" Mutant of Temperate Phage P 2. Bact. Proc. 10, 38—39.
— 1958: Lysogeny. Adv. Virus Res. 5 (in press).

BERTANI, L. E., 1957: The Effect of the Inhibition of Protein Synthesis on the Establishment of Lysogeny. Virology 4, 53—71.

Best, R. J., and H. P. C. Gallus, 1955: Further Evidence for the Transfer of Character-Determinants (Recombination) between Strains of Tomato Spotted Wilt Virus. Enzymologia **17**, 207—221.
Beveridge, W. I. B., and F. M. Burnet, 1946: The Cultivation of Viruses and Rickettsiae in the Chick Embryo. Med. Res. Counc. Rep. Series No. **256**.
Bird, F. T., 1957: On the Development of Insect Viruses. Virology **3**, 237—242.
Black, L. M., 1953: Viruses that Reproduce in Plants and Insects. Ann. N.Y. Acad. Sci. **56**, 398—413.
— and M. K. Brakke, 1954: Serological Reactions of a Plant Virus Transmitted by Leafhopper. Phytopath. **44**, 282.
Bland, J. O. W., and C. F. Robinow, 1939: The Inclusion Bodies of Vaccinia and their Relationship to the Elementary Bodies Studied in Cultures of the Rabbit's Cornea. J. Path. a. Bacter. **48**, 381—403.
Brenner, S., 1957: Genetic Control and Phenotypic Mixing of the Adsorption Cofactor Requirement in Bacteriophages T 2 and T 4. Virology **3**, 560—574.
Bresch, C., 1955: Zum Paarungsmechanismus von Bakteriophagen. Z. Naturforsch. **10 b**, 545—561.
— and T. Trautner, 1955: Zur Kinetik der Rekombinantenbildung bei T 1-Bakteriophagen. Z. Naturforsch. **10 b**, 436—440.
Brown, G. L., and N. Symonds, 1951: Chemical Basis of Heredity. Discussion of paper by Levinthal and Thomas (see reference).
Burnet, F. M., 1951: Mucoproteins in Relation to Virus Action. Physiol. Rev. (Am.) **31**, 131—150.
— 1955: Principles of Animal Virology. New York.
— 1956: Structure of Influenza Virus. Science **123**, 1101—1104.
— and J. O. Stone, 1947: The Receptor Destroying Enzyme of *V. cholerae*. Austral. J. exper. Biol. a. Med. Sci. **25**, 227—233.
Cairns, H. J. F., 1957: The Asynchrony of Infection by Influenza Virus. Virology **3**, 1—14.
— and P. J. Mason, 1953: Production of Influenza A Virus in the Cells of the Allantois. J. Immunol. **71**, 38—40.
Chambers, V. C., 1957: The Prolonged Persistence of Western Equine Encephalomyelitis Virus in Cultures of Strain L Cells. Virology **3**, 62—75.
Chase, M., and A. D. Doermann, 1958: High Negative Interference over Short Segments of the Genetic Structure of Bacteriophage T 4. Genetics (in press).
Cieciura, S. J., P. I. Marcus, and T. T. Puck, 1957: The Use of X-Irradiated HeLa Cell Giants to Detect Latent Virus in Mammalian Cells. Virology **3**, 426—428.
Clark, P. F., 1949: The Influence of Nutrition on Experimental Virus Infection. Bacter. Rev. **13**, 122—127.
Cohen, S. S., 1949: Growth Requirements of Bacterial Viruses. Bacter. Rev. **13**, 1—24.
— 1953: Studies on Controlling Mechanisms in the Metabolism of Virus-Infected Bacteria. Cold Spring Harbor Symposia Quant. Biol. **18**, 221—235.
Colter, J. S., H. H. Bird, R. A. Brown, 1957: Infectivity of Ribonucleic Acid from Ehrlich Ascites Tumour Cells Infected with Mengo Encephalitis. Nature **179**, 859—860.
Commoner, B., and P. M. Dietz, 1952: Changes in Non-Protein Nitrogen Metabolism during Tobacco Mosaic Virus Biosynthesis. J. gen. Physiol. (Am.) **35**, 847—856.
— J. A. Lippincott, G. B. Shearer, E. E. Richman, and J. H. Wu, 1956: Reconstitution of Tobacco Mosaic Virus Components. Nature **178**, 767—771.
— and F. L. Mercer, 1952: The Effect of Thiouracil on the Rate of Tobacco Mosaic Virus Biosynthesis. Arch. Biochem. Biophysics **35**, 278—289.
— — P. Merrill, and A. J. Zimmer, 1950: Microanalytical Determination of the Rate of Tobacco Mosaic Virus Synthesis in Tobacco Leaf Tissue. Arch. Biochemistry **27**, 271—286.
— P. Newmark, and S. D. Rodenberg, 1952: An Electrophoretic Analysis of Tobacco Mosaic Virus Biosynthesis. Arch. Biochem. Biophysics **37**, 15—36.
— D. L. Scheiber, and P. M. Dietz, 1953: Relationships between Tobacco Mosaic Virus Biosynthesis and the Nitrogen Metabolism of the Host. J. gen. Physiol. (Am.) **36**, 807—830.
Coons, A. H., 1957: The Morphological Aspects of Virus Infections of Cells as Revealed by Fluorescent Antibody. In: The Nature of Viruses. Ciba Foundation Symposium, London.
Crick, F. H. C., and J. D. Watson, 1957: Virus Structure: General Principles. In: The Nature of Viruses. Ciba Foundation Symposium, London.

Delbrück M., and W. T. Bailey Jr., 1946: Induced Mutations in Bacterial Viruses. Cold Spring Harbor Symposia Quant. Biol. **11**, 33—37.
— and G. S. Stent, 1957: On the Mechanism of DNA Replication. In: McElroy and Glass, The Chemical Basis of Heredity. Baltimore.
Delwiche, C. C., P. Newmark, W. N. Takahashi, and M. J. Ng, 1955: The Relationship between Tobacco Mosaic Virus and an Accompanying Abnormal Protein. Biochim. Biophys. Acta **16**, 127—136.
DeMars, R. I., 1955: The Production of Phage-Related Materials when Bacteriophage Development is Interrupted by Proflavine. Virology **1**, 83—99.
Doermann, A. H., 1952: The Intracellular Growth of Bacteriophage. I. Liberation of Intracellular Bacteriophage T 4 by Premature Lysis with Another Phage or with Cyanide. J. gen. Physiol. (Am.) **35**, 645—656.
— 1953: The Vegetative State in the Life Cycle of Bacteriophage: Evidence for its Occurrence and its Genetic Characterization. Cold Spring Harbor Symposia Quant. Biol. **18**, 3—11.
— M. Chase, and F. W. Stahl, 1955: Genetic Recombination and Replication in Bacteriophage. J. cellul. a. comp. Physiol. (Am.) **45** (Suppl. 2), 51—74.
Donald, H. B., and A. Isaacs, 1954: Some Properties of Influenza Virus Filaments Shown by Electron Microscopic Particle Counts. J. gen. Microbiol. **11**, 325—331.
Dulbecco, R., 1950: Experiments on Photoreactivation of Bacteriophages Inactivated with Ultraviolet Radiation. J. Bacter. (Am.) **59**, 329—347.
— 1952: Production of Plaques in Monolayer Tissue Cultures by Single Particles of an Animal Virus. Proc. Nat. Acad. Sci. **38**, 747—752.
— 1957: Quantitative Aspects of Virus Growth in Cultivated Animal Cells. In: The Nature of Viruses. Ciba Foundation Symposium, London.
— and M. Vogt, 1953: Some Problems of Animal Virology as Studied by the Plaque Technique. Cold Spring Harbor Symposia Quant. Biol. **18**, 273—279.
— — 1954: One Step Growth Curve of Western Equine Encephalomyelitis Virus on Chicken Embryo Cells Grown *in Vitro* and Analysis of Virus Yields from Single Cell. J. exper. Med. (Am.) **99**, 183—199.
— — 1955: Biological Properties of Poliomyelitis Viruses as Studied by the Plaque Technique. Ann. N. Y. Acad. Sci. **61**, 790—800.
Dunn, D. B., and J. D. Smith 1954: Incorporation of Halogenated Pyrimidines into the Desoxyribonucleic Acids of Bacterium coli and its Bacteriophage. Nature **174**, 305—306.
Duran-Reynals, F., 1940: A Hemorrhagic Disease Occurring in Chicks Inoculated with Rous and Fujinami Viruses. Yale J. Biol. Med. **13**, 77—98.
Eagle, H., and K. Habel, 1956: The Nutritional Requirements for the Propagation of Poliomyelitis Virus by the HeLa Cell. J. exper. Med. (Am.) **104**, 271—287.
Edgar, R., 1956: Mutation and Recombination of the Host Range Genetic Region of Phage T 2. Quoted by: Streisinger and Franklin, Cold Spring Harbor Symposia Quant. Biol. **21**, 103—111.
Enders, J. F., 1949: Remarks on Cellular Resistance to Mammalian Viruses. Feder. Proc. **8**, 625—630.
Ephrussi, B., 1953: Nucleo-Cytoplasmic Relations in Micro-Organisms. Oxford.
Fazekas de St. Groth S., 1948: Viropexis, the Mechanism of Influenza Virus Infection. Nature **162**, 294—295.
Flewett, T. H., and C. E. Challice, 1951: The Intracellular Growth of Fowl-Plague Virus; a Phase-Contrast and Electron Microscopical Study of Infected Tissue Cultures. J. gen. Microbiol. **5**, 279—286.
Fraenkel-Conrat, H., 1956: The Role of the Nucleic Acid in the Reconstitution of Active Tobacco Mosaic Virus. J. Amer. Chem. Soc. **78**, 882—883.
— and R. C. Williams, 1955: Reconstitution of Active Tobacco Mosaic Virus from its Inactive Protein and Nucleic Acid Components. Proc. Nat. Acad. Sci. **41**, 690—698.
Franklin, R. E., A. Klug, and K. C. Holmes, 1957: X-ray Diffraction Studies of the Structure and Morphology of Tobacco Mosaic Virus. In: The Nature of Viruses. Ciba Foundation Symposium, London.
Franklin, R. M., H. Rubin, and C. A. Davis, 1957: The Production, Purification, and Properties of Newcastle Disease Virus Labeled with Radiophosphorus. Virology **3**, 96—114.
Friedkin, M., 1954: Enzymatic Synthesis of Azaguanine Riboside and Azaguanine Desoxyriboside. J. Biol. Chem. **209**, 295—301.
Frisch-Niggemeyer, W., 1956: Effect of Ionizing Radiations on Serum Complement and Enzymes. Nature **178**, 307—308.

Frisch-Niggemeyer, W., and L. Hoyle, 1956: The Nucleic Acid and Carbohydrate Content of Influenza Virus A and of Virus Fractions Produced by Ether Disintegration. J. Hyg. **54**, 201—212.
Fulton, F., and A. Isaacs, 1953: Influenza Virus Multiplication in the Chick Chorioallantoic Membrane. J. gen. Microbiol. **9**, 119—131.
Garen, A., and N. D. Zinder, 1955: Radiological Evidence for Partial Genetic Homology between Bacteriophage and Host Bacteria. Virology **1**, 347—376.
Gaylord, W. H., and J. L. Melnick, 1953: Intracellular Forms of Pox Viruses as Shown by the Electron Microscope (Vaccinia, Ectromelia, Molluscum Contagiosum). J. exper. Med. (Am.) **98**, 157—172.
Gey, G. O., and F. B. Bang, 1951: Viruses and Cells—A Study in Tissue Culture Applications. Trans. New York Acad. Sci. **14**, 15—24.
Gierer, A., and G. Schramm, 1956: Die Infektiosität der Nucleinsäure aus Tabakmosaikvirus. Z. Naturforsch. **11 b**, 138—142.
Ginoza, W., and A. Norman, 1957: Radiosensitive Molecular Weight of Tobacco Mosaic Virus Nucleic Acid. Nature **179**, 520—521.
Gostling, J. V. T., and S. P. Bedson, 1956: Observations on the Mode of Multiplication of Herpes Virus. Brit. J. exper. Path. **37**, 434—446.
Gottschalk, A., 1956: Neuraminic Acid: The Functional Group of Some Biologically Active Mucoproteins. Yale J. Biol. Med. **28**, 525—537.
— 1957: Virus Enzymes and Virus Templates. Physiol. Rev. (Am.) **37**, 66—83.
Granoff, A., and G. K. Hirst, 1954: Experimental Production of Combination Forms of Virus. IV. Mixed Influenza A-Newcastle Disease Virus Infection. Proc. Soc. exper. Biol. a. Med. (Am.) **86**, 84—88.
Groman, N. B., 1955: Evidence for the Active Role of Bacteriophage in the Conversion of Nontoxigenic *Corynebacterium diphtheriae* to Toxin Production. J. Bacter. (Am.) **69**, 9—15.
Haas, V. H., S. E. Stewart, and G. M. Briggs, 1957: Folic Acid Deficiency and the Sparing of Mice Infected with the Virus of Lymphocytic Choriomeningitis. Virology **3**, 15—21.
Hamers-Casterman, C., and R. Jeener, 1957: An Initial Ribonuclease-Sensitive Phase in the Multiplication of Tobacco Mosaic Virus. Virology **3**, 197—206.
Henle, G., A. Girardi, and W. Henle, 1955: A Non-transmissible Cytopathogenic Effect of Influenza Virus in Tissue Culture Accompanied by Formation of Non-infectious Hemagglutinins. J. exper. Med. (Am.) **101**, 25—41.
Henle, W., 1950: Interference Phenomenon between Animal Viruses: A Review. J. Immunol. **64**, 203—236.
— 1953: Multiplication of Influenza Virus in the Entodermal Cells of the Allantois of the Chick Embryo. Adv. Virus Res. **1**, 141—227.
Hershey, A. D., 1953: Nucleic Acid Economy in Bacteria Infected with Bacteriophage T 2. II. Phage Precursor Nucleic Acid. J. Gen. Physiol. (Am.) **37**, 1—23.
— 1956: Chemistry and Viral Growth. In: Green, Currents in Biochemical Research. New York.
— 1957: Bacteriophages as Genetic and Biochemical Systems. Adv. Virus Res. **4**, 25—61.
— and E. Burgi, 1956: Genetic Significance of the Transfer of Nucleic Acid from Parental to Offspring Phage. Cold Spring Harbor Symposia Quant. Biol. **21**, 91—101.
— and M. Chase, 1951: Genetic Recombination and Heterozygosis in Bacteriophage. Cold Spring Harbor Symposia Quant. Biol. **16**, 471—479.
— — 1952: Independent Functions of Viral Protein and Nucleic Acid in Growth of Bacteriophage. J. gen. Physiol. (Am.) **36**, 39—56.
— M. D. Kamen, J. W. Kennedy, and H. Gest, 1951: The Mortality of Bacteriophage Containing Assimilated Radioactive Phosphorus. J. gen. Physiol. (Am.) **34**. 305—319.
— and N. E. Melechen, 1957: Synthesis of Phage-Precursor Nucleic Acid in the Presence of Chloramphenicol. Virology **3**, 207—236.
— and R. Rotman, 1949: Genetic Recombination between Host-Range and Plaque-Type Mutants of Bacteriophage in Single Bacterial Cells. Genetics **34**, 44—71.
Hirst, G. K., 1941: The Agglutination of Red Cells by Allantoic Fluid of Chick Embryos Infected with Influenza Virus. Science **94**, 22—23.
— 1953: Intracellular Reactions between Two Types of Influenza Virus. Cold Spring Harbor Symposia Quant. Biol. **18**, 25—27.
— T. Gotlieb, and A. Granoff, 1957: Studies on Mixed Infections with Influenza Viruses. In: The Nature of Viruses. Ciba Foundation Symposium, London.

HOYLE, L., 1948: The Growth Cycle of Influenza Virus A. A Study of the Relations between Virus, Soluble Antigen and Host Cell in Fertile Eggs Inoculated with Influenza Virus. Brit. J. exper. Path. **29**, 390—399.

— 1952: The Multiplication of Complement Fixing Antigen and Red-Cell Agglutination in the Chorio-Allantoic Membrane of Fertile Eggs Inoculated with Influenza Virus. J. Path. a. Bacter. **64**, 419—423.

— 1952: Structure of the Influenza Virus: The Relation between Biological Activity and Chemical Structure of Virus Fractions. J. Hyg. **50**, 229—245.

— 1957: The Use of Radioactive Influenza Virus to Determine the Fate of the Infecting Particle on Entry into the Host Cell. In: The Nature of Viruses. Ciba Foundation Symposium, London.

HURST, E. W., and R. HULL, 1956: The Chemotherapy of Virus Disease, with Brief Consideration of the Influence of Dietary, Hormonal and Other Factors in Virus Infections. Pharmacol. Rev. **8**, 199—263.

ISAACS, A., and F. FULTON, 1953: Interference in the Chick Chorion. J. gen. Microbiol. **9**, 132—139.

— 1957: Particle Counts and Infectivity Titrations for Animal Viruses. Adv. Virus Res. **4**, 111—158.

JACOB, F., 1954: Les Bactéries Lysogènes et la Notion de Provirus. Paris.

— et E. L. WOLLMAN, 1955: Étude Génétique d'un Bactériophage Tempéré d'*Escherichia coli.* III. Effet du Rayonnement Ultraviolet sur la Recombinaison Génétique. Ann. Inst. Pasteur **88**, 724—749.

— — 1956: Recherches sur les Bactéries Lysogènes Defectives. I. Determinisme Génétique de la Morphogenèse chez un Bactériophage Tempéré. Ann. Inst. Pasteur **90**, 282—302.

— 1957: Genetic Aspects of Lysogeny. In: MCELROY and GLASS, The Chemical Basis of Heredity. Baltimore.

JEENER, R., 1954: Influence of Thiouracil Incorporation in the Ribonucleic Acid Moiety of Tobacco Mosaic Virus on its Multiplication. Biochem. Biophys. Acta **13**, 148—149.

— 1956: Ribonucleic Acids and Virus Multiplication. Adv. Enzymol. **17**, 477—498.

— 1957: Biological Effects of the Incorporation of Thiuracil into the Ribonucleic Acid of Tobacco Mosaic Virus. Biochem. Biophys. Acta **23**, 351—361.

— and P. LEMOINE, 1953: Occurrence in Plants Infected with Tobacco Mosaic Virus of a Crystallizable Antigen Devoid of Ribonucleic Acid. Nature **171**, 935—936.

— and C. VAN RYSSELBERGE, 1955: Un Essai d'Infection Directe des Cellules du Parenchyme de la Feuille de Tabac à l'Aide de Virus de la Mosaique Marqué par le ^{14}C et le ^{32}P. Sort du Virus. Biochem. Biophys. Acta **17**, 233—239.

KAISER, A. D., 1955: A Genetic Study of the Temperature Coliphage λ. Virology **1**, 424—443.

— 1957: Mutations in a Temperate Bacteriophage Affecting its Ability to Lysogenize *Escherichia coli.* Virology **3**, 42—61.

KAPLAN, A. S., 1955: The Susceptibility of Monkey Kidney Cells to Poliovirus *in vivo* and *in vitro.* Virology **1**, 377—392.

KELLENBERGER, E., and W. ARBER, 1955: Die Struktur des Schwanzes der Phagen T 2 and T 4 und der Mechanismus der irreversiblen Adsorption. Z. Naturforsch. **10 b**, 698—704.

— and J. SÉCHAUD, 1957: Electron Microscopical Studies of Phage Multiplication. II. Production of Phage-Related Structures during Multiplication of Phages T 2 and T 4. Virology **3**, 256—274.

KNIGHT, C. A., 1946: Precipitin Reactions of Highly Purified Influenza Viruses and Related Materials. J. exper. Med. (Am.) **83**, 281—294.

— 1954: The Chemical Constitution of Viruses. Adv. Virus Res. **2**. 153—182.

— 1957: Some Recent Developments in the Chemistry of Virus Mutants. In: The Nature of Viruses. Ciba Foundation Symposium, London.

KORNBERG, A., 1957: Pathways of Enzymatic Synthesis of Nucleotides and Polynucleotides. In: MCELROY and GLASS, The Chemical Basis of Heredity, Baltimore.

KOZLOFF, L. M., 1953: Origin and Fate of Bacteriophage Material. Cold Spring Harbor Symposia Quant. Biol. **18**, 209—220.

— M. LUTE, and K. HENDERSON, 1957: Viral Invasion. I. II. III. J. biol. Chem. (Am.) **228**, 511—546.

LANNI, F., and Y. T. LANNI, 1953: Antigenic Structure of Bacteriophage. Cold Spring Harbor Symposia Quant. Biol. **18**, 159—168.

LEBRUN, J., 1956: Cellular Localization of Herpes Simplex Virus by Means of Fluorescent Antibody. Virology **2**, 496—510.

LeClerc, J., 1956: Action of Ribonuclease on the Multiplication of the Influenza Virus. Nature **177**, 578—579.
Lederberg, E. M., and J. Lederberg, 1953: Genetic Studies of Lysogenicity in *Escherichia coli*. Genetics **38**, 51—64.
Lederberg, J., 1952: Cell Genetics and Hereditary Symbiosis. Physiol. Rev. (Am.) **32**, 403—430.
— 1955: Recombination Mechanisms in Bacteria. J. cellul. a. comp. Physiol. (Am.) **45** (Suppl. 2), 75—107.
Lennox, E. S., 1955: Transduction of Linked Genetic Characters of the Host by Bacteriophage P 1. Virology **1**, 190—206.
Levine, M., 1957: Mutations in the Temperate Phage P 22 and Lysogeny in *Salmonella*. Virology **3**, 22—41.
Levine, S., T. T. Puck, and B. P. Sagik, 1953: An Absolute Method for Assay of Virus Hemagglutinins. J. exper. Med. (Am.) **98**, 521—531.
— and B. P. Sagik, 1956: The Interactions of Newcastle Disease Virus (NDV) with Chick Embryo Tissue Culture Cells: Attachment and Growth. Virology **2**, 57—68.
Levinthal, C., 1954: Recombination in Phage T 2: Its Relationship to Heterozygosis and Growth. Genetics **39**, 169—184.
— 1956: The Mechanism of DNA Replication and Genetic Recombination in Phage. Proc. Nat. Acad. Sci. **32**, 394—404.
— and H. R. Crane, 1956: On the Unwinding of DNA. Proc. Nat. Acad. Sci. **42**, 436—438.
— and C. A. Thomas Jr., 1957: The Molecular Basis of Genetic Recombination in Phage. In: McElroy and Glass, The Chemical Basis of Heredity, Baltimore.
L'Héritier, Ph., 1951: The CO_2 Sensitivity Problem in *Drosophila*. Cold Spring Harbor Symposia Quant. Biol. **16**, 99—112.
— 1957: The Hereditary Virus of *Drosophila*. Adv. Virus Res. **5** (in press).
Lieb, M., 1953: The Establishment of Lysogenicity in *Escherichia coli*. J. Bacter. (Am.) **65**, 642—651.
Lief, F. S., and W. Henle, 1956: Studies on the Soluble Antigen of Influenza Virus. I. The Release of S Antigen from Elementary Bodies by Treatment with Ether. Virology **2**, 753—771.
Lind, P. E., and F. M. Burnet, 1953: Back-Recombination of Influenza A Strains Obtained in Recombination Experiments. Austral. J. exper. Biol. a. med. Sci. **31**, 361—372.
Lindegren, C. C., 1955: Non-Mendelian Segregation in a Single Tetrad of Saccharomyces Ascribed to Gene Conversion. Science **121**, 605—607.
Litman, R. M., and A. B. Pardee, 1956: Production of Bacteriophage Mutants by a Disturbance of Desoxyribonucleic Acid Metabolism. Nature **178**, 529—531.
Liu, C., 1955: Studies on Influenza Infection in Ferrets by Means of Fluorescein-Labelled Antibody. I. II. J. exper. Med. (Am.) **101**, 665—686.
Loh, P. C., H. F. Maassab, and W. W. Ackermann, 1957: Phosphorus Metabolism of HeLa Cells Infected with Poliovirus. Bacter. Proc. **138**.
Luria, S. E., 1951: The Frequency Distribution of Spontaneous Bacteriophage Mutants as Evidence for the Exponential Rate of Phage Reproduction. Cold Spring Harbor Symposia Quant. Biol. **16**, 463—470.
— 1953 a: Host-Induced Modifications of Viruses. Cold Spring Harbor Symposia Quant. Biol. **18**, 237—244.
— 1953 b: General Virology. New York.
— and M. L. Human, 1950: Chromation Staining of Bacteria during Bacteriophage Infection. J. Bacter. **59**, 551—560.
— and D. L. Steiner, 1954: The Role of Calcium in the Penetration of Bacteriophage T 5 into its Host. J. Bacter. (Am.) **67**, 635—639.
— R. C. Williams, and R. C. Backus, 1951: Electron Micrographic Counts of Bacteriophage Particles. J. Bacter. (Am.) **61**, 179—188.
Lwoff, A., 1953: Lysogeny. Bacter. Rev. **17**, 269—337.
— R. Dulbecco, M. Vogt, and M. Lwoff, 1955: Kinetics of the Release of Poliomyelitis Virus from Single Cells. Virology **1**, 128—139.
Maaløe, O., and J. D. Watson, 1951: The Transfer of Radioactive Phosphorus from Parental to Progeny Phage. Proc. Nat. Acad. Sci. **37**, 507—513.
Maramorosch, K., 1955: Multiplication of Plant Viruses in Insect Vectors. Adv. Virus Res. **3**, 221—249.
Markham, R., and K. M. Smith, 1949: Studies on the Virus of Turnip Yellow Mosaic. Parasitology **39**, 330—342.

MATTHEWS, R. E. F., and J. D. SMITH, 1955: The Chemotherapy of Viruses. Adv. Virus Res. **3**, 49—148.
MENEGHINI, M., and C. C. DELWICHE, 1951: The Multiplication of Tobacco Mosaic Virus in the Host Plant. J. Biol. Chem. **189**, 177—186.
MERRILL, M. H., and C. TENBROECK, 1935: The Transmission of Equine Encephalomyelitis Virus by *Aedes aegypti*. J. exper. Med. (Am.) **62**, 687—695.
MORGAN, C., G. H. BERGOLD, D. H. MOORE, and H. M. ROSE, 1955: The Macromolecular Paracrystalline Lattice of Insect Viral Polyhedral Bodies Demonstrated in Ultrathin Sections Examined in the Electron Microscope. J. Biophys. Biochem. Cytol. **1**, 187—190.
— S. A. ELLISON, H. M. ROSE, and D. H. MOORE, 1954a: Structure and Development of Viruses as Observed in the Electron Microscope. I. Herpes Simplex Virus. J. exper. Med. (Am.) **100**, 195—202.
— — — — 1954b: Structure and Development of Viruses Observed in the Electron Microscope. II. Vaccinia and Fowl Pox Viruses. J. exper. Med. (Am.) **100**, 301—310.
— C. HOWE, H. M. ROSE, and D. H. MOORE, 1956: Structure and Development of Viruses Observed in the Electron Microscope. IV. Viruses of the RI-APC Group. J. Biophys. Biochem. Cytol. **2**, 351—360.
— H. M. ROSE, and D. H. MOORE, 1956: Structure and Development of Viruses Observed in the Electron Microscope. III. Influenza Virus. J. exper. Med. (Am.) **104**, 171—182.
MORSE, M. L., E. M. LEDERBERG, and J. LEDERBERG, 1956a: Transduction in *Escherichia coli* K-12. Genetics **41**, 142—156.
— — — 1956b: Transduction Heterogenotes in *Escherichia coli.* Genetics **41**, 758—779.
MOULDER, J. W., 1954: Biochemical Aspects of the Growth of Feline Pneumonitis Virus in the Chick Embryo Yolk Sac. Bacter. Rev. **18**, 170—176.
NIXON, H. L., 1956: An Estimate of the Number of Tobacco Mosaic Virus Particles in a Single Hair Cell. Virology **2**, 126—128.
OBERLING, C., 1952: The Riddle of Cancer. New Haven, Conn.
OVERMAN, J. R., and I. TAMM, 1957: Multiplication of Vaccinia Virus in the Chorioallantoic Membrane *in vitro*. Virology **3**, 173—184.
PALADE, G. E., 1956: The Endoplasmic Reticulum. J. Biophys. Biochem. Cytol. **2** (Suppl.), 85—98.
PRICE, W. C., 1940: Acquired Immunity from Plant Virus Diseases. Quart. Rev. Biol. **15**, 338—361.
PUCK, T. T., 1957: The Mammalian Cell as Microorganism. In: Symp. Society for the Study of Development and Growth. Princeton, New Jersey.
— and P. I. MARCUS, 1956: Action of X-Rays on Mammalian Cells. J. exper. Med. (Am.) **103**, 653—666.
— — and S. J. CIECIURA, 1956: Clonal Growth of Mammalian Cells *in vitro*. Growth Characteristics of Colonies from Single HeLa Cells with and without a "Feeder" Layer. J. exper. Med. (Am.) **103**, 273—284.
RAPPAPORT, C., 1956: Monolayer Cultures of Trypsinized Monkey Kidney Cells in Synthetic Medium. Application to Poliovirus Synthesis. Proc. Soc. exper. Biol. a. Med. (Am.) **91**, 464—470.
ROMAN, H., 1956: Studies of Gene Mutation in Saccharomyces. Cold Spring Harbor Symposia Quant. Biol. **21**, 175—185.
ROUS, P., 1946: Concerning the Cancer Problem. Amer. Scientist **34**, 329—357.
RUBIN, H., 1955: Quantitative Relations between Causative Virus and Cell in the Rous No. I Chicken Sarcoma. Virology **1**, 445—473.
— M. BALUDA, and J. E. HOTCHIN, 1955: The Maturation of Western Equine Encephalomyelitis Virus and its Release from Chick Embryo Cells in Suspension. J. exper. Med. (Am.) **101**, 205—212.
— and R. M. FRANKLIN, 1957: On the Mechanism of Newcastle Disease Virus Neutralization by Immune Serum. Virology **3**, 84—95.
— — and M. BALUDA, 1957: Infection and Growth of Newcastle Disease Virus (NDV) in Cultures of Chick Embryo Lung Epithelium. Virology **3**, 587—600.
SANDERS, F. K., 1957: The Multiplication of Animal Viruses. In: The Nature of Viruses. Ciba Foundation Symposium, London.
SCHÄFER, W., 1957: Units Isolated After Splitting Fowl Plague Virus. In: The Nature of Viruses. Ciba Foundation Symposium, London.
SCHLESINGER, R. W., 1953: Developmental Stages of Virus. Ann. Rev. Microbiol. **7**, 83—112.

Schramm, G., 1947 a: Über die Spaltung des Tabakmosaikvirus und die Wiedervereinigung der Spaltstücke zu höhermolekularen Proteinen. I. Die Spaltungsreaktion. Z. Naturforsch. **2 b**, 112—121.
— 1947 b: Über die Spaltung des Tabakmosaikvirus und die Wiedervereinigung der Spaltstücke zu höhermolekularen Proteinen. II. Versuche zur Wiedervereinigung der Spaltstücke. Z. Naturforsch. **2 b**, 249—257.
Scott, T. F. McN., L. L. Coriell, H. Blank, and A. Gray, 1953: The Growth Curve of the Virus of Herpes Simplex on the Chorioallantoic Membrane of the Embryonated Hen's Egg. J. Immunol. **71**, 134—144.
Séchaud, J., and E. Kellenberger 1956: Lyse Precoce, Provoquée par le Chloroforme chez les Bactéries Infectées par du Bactériophage. Ann. Inst. Pasteur **90**, 102—106.
Siegel, A., W. Ginoza, and S. G. Wildman, 1957: The Early Events of Infection with Tobacco Mosaic Virus Nucleic Acid. Virology **3**, 554—559.
— and S. G. Wildman, 1956: The Inactivation of the Infectius Centers of Tobacco Mosaic Virus by Ultraviolet Light. Virology **2**, 69—82.
— — and W. Ginoza, 1956: Sensitivity to Ultraviolet Light of Infectious Tobacco Mosaic Virus Nucleic Acid. Nature **178**, 1117—1118.
Sigel, M. M., A. J. Girardi, and E. G. Allen, 1951: Studies on the Psittacosis-Lymphogranuloma Group. J. exper. Med. (Am.) **94**, 401—413.
Simkin, J. L., and T. S. Work, 1957: Biochemical Approaches to the Problem of Protein Synthesis. Nature **179**, 1214—1219.
Smadel, J. E., and C. L. Hoagland, 1942: Elementary Bodies of Vaccinia. Bacter. Rev. **6**, 79—110.
Smith, K. M., 1955: Morphology and Development of Insect Viruses. Adv. Virus Res. **3**, 199—220.
Smith, W., G. Belyavin and F. W. Scheffield, 1955: The Host-Tissue Component of Influenza Viruses. Proc. Roy. Soc. **B.**, **143**, 504—522.
Sprunt, K., I. M. Mountain, W. M. Redman, and H. E. Alexander, 1955: Production of Poliomyelitis Virus with Combined Antigenic Characteristics of Type I and Type II. Virology **1**, 236—249.
Stahl, F. W., 1956: The Effects of the Decay of Incorporated Radioactive Phosphorus on the Genome of Bacteriophage T 4. Virology **2**, 206—234.
Steere, R. L., 1952: Virus Increment Curve Obtained from Counts of Particles in Clarified Plant Juice. Amer. J. Bot. **39**, 211—220.
— and R. C. Williams, 1953: Identification of Crystalline Inclusion Bodies Extracted Intact from Plant Cells Infected with Tobacco Mosaic Virus. Amer. J. Bot. **40**, 81—84.
Stent, G. S., 1955: Decay of Incorporated Radioactive Phosphorus during Reproduction of Bacteriophage T 2. J. gen. Physiol. (Am.) **38**, 853—865.
— 1958: Mating in the Reproduction of Bacterial Viruses. Adv. Virus Res. **5** (in press).
— C. R. Fuerst, and F. Jacob, 1957: Inactivation d'un Prophage par la Désintegration du Radiophosphore. C. r. Acad. Sci. **244**, 1840—1842.
— and E. L. Wollman, 1950: Studies on Activation of T 4 Bacteriophage by Cofactor. II. The Mechanism of Activation. Biochem. Biophys. Acta **6**, 307—316.
Streisinger, G., 1956 a: The Genetic Control of Host Range and Serological Specificity in Bacteriophages T 2 and T 4. Virology **2**, 377—387.
— 1956 b: Phenotypic Mixing of Host Range and Serological Specificities in Bacteriophages T 2 and T 4. Virology **2**, 388—398.
— and N. C. Franklin, 1956: Mutation and Recombination at the Host Range Genetic Region of Phage T 2. Cold Spring Harbor Symposia Quant. Biol. **21**, 103—111.
Takahashi, W. N., and M. Ishii, 1953: A Macromolecular Protein Associated with Tobacco Mosaic Virus Infection: Its Isolation and Properties. Amer. J. Bot. **40**, 85—90.
Tamm, I., 1955: Selective Inhibition of Virus Multiplication by Synthetic Chemicals. Bull. N.Y. Acad. Med. **31**, 537—540.
Taylor, J. H., 1957: The Time and Mode of Duplication of Chromosomes. Amer. Nat. **91**, 209—221.
— P. S. Woods, and W. L. Hughes, 1956: The Organization and Duplication of Chromosomes as Revealed by Autoradiographic Studies Using Tritium-Labeled Thymidine. Proc. Nat. Acad. Sci. **43**, 122—128.
Tolmach, L. J., 1957: Attachment and Penetration of Cells by Viruses. Adv. Virus Res. **4**, 63—110.

TOMIZAWA, J., and S. SUNAKAWA, 1956: The Effect of Chloramphenicol on Desoxyribonucleic Acid Synthesis and the Development of Resistance to Ultraviolet Irradiation in *E. coli* Infected with Bacteriophage T 2. J. gen. Physiol. (Am.) **39**, 553—565.
TRAUB, E., 1939: Epidemiology of Lymphocytic Choriomeningitis in a Mouse Stock Observed for Four Years. J. exper. Med. (Am.) **69**, 801—817.
UETAKE, H., S. E. LURIA, and J. W. BURROUS, 1958: Conversion of Somatic Antigens in Salmonella by Phage Infection Leading to Lysis or Lysogeny. Virology **5** 68—91.
— T. NAKAGAWA, and T. AKIBA. 1955: The Relationship of Bacteriophage to Antigenic Changes in Group E Salmonellas. J. Bacter. (Am.) **69**, 571—579.
VAGO, C., 1951: The Phenomenon of Latency in a Disease Caused by the Ultravirus of Insects. Rev. Canad. Biol. **10**, 299—307.
VAN RYSSELBERGE, C., and R. JEENER, 1957: Plant Virus Synthesis and the Abnormal Protein Constituents of Infected Leaves. Biochem. Biophys. Acta **23**, 18—23.
VISCONTI, N., and M. DELBRÜCK, 1953: The Mechanism of Genetic Recombination in Phage. Genetics **38**, 5—33.
VOGT, M., R. DULBECCO, and H. A. WENNER, 1957: Mutants of Poliomyelitis Viruses with Reduced Efficiency of Plating in Acid Medium and Reduced Neuropathogenicity. Virology **4**, 141—155.
VOLKIN, E., and L. ASTRACHAN, 1957: RNA Metabolism in T 2-Infected *Escherichia coli.* In: MCELROY and GLASS, The Chemical Basis of Heredity. Baltimore.
VON MAGNUS, P., 1954: Incomplete Forms of Influenza Virus. Adv. Virus Res. **2**, 59—79.
WANG, T. Y., and B. COMMONER, 1956: The Formation of Infectious Nucleoprotein from Tobacco Mosaic Virus Protein and Tobacco Leaf DNA. Proc. Nat. Acad. Sci. **42**, 831—841.
WATSON, J. D., and F. H. C. CRICK, 1953: The Structure of DNA. Cold Spring Harbor Symposia Quant. Biol. **18**, 123—131.
WEIGLE, J. J., 1953: Induction of Mutations in a Bacterial Virus. Proc. Nat. Acad. Sci. **39**, 628—636.
WHITFIELD J. F., and R. G. E. MURRAY, 1954: A Cytological Study of the Lysogenization of *Shigella dysenteriae* by P 1 and P 2 Bacteriophages. Canad. J. Microbiol. **1**, 216—226.
WILDY, P., 1954: The Growth of Herpes Simplex Virus. Austral. J. exper. Biol. a. Med. Sci. **32**, 605—619.
— 1955: Recombination with Herpes Simplex Virus. J. gen. Microbiol. **13**, 346—360.
WILLIAMS, R. C., 1953: The Shapes and Sizes of Purified Viruses as Determined by Electron Microscopy. Cold Spring Harbor Symposia Quant. Biol. **18**, 185—195.
— 1957: Structure and Substructure of Viruses as Seen Under the Electron Microscope. In: The Nature of Viruses. Ciba Foundation Symposium, London.
— and K. M. SMITH, 1957: A Crystallizable Insect Virus. Nature **179**, 119—120.
WITHMAN, L., 1937: The Multiplication of the Virus of Yellow Fever in *Aedes aegypti.* J. exper. Med. (Am.) **66**, 133—143.
WOLLMAN, E. L., F. JACOB and W. HAYES, 1956: Conjugation and Genetic Recombination in *Escherichia coli* K-12. Cold Spring Harbor Symposia Quant. Biol. **21**, 141—162.
WYATT, G. R., and S. S. COHEN, 1952: A New Pyrimidine Base From Bacteriophage Nucleic Acids. Nature **170**, 1072—1073.
XEROS, N., 1955: Origin of the Virus-Producing Chromatic Mass or Net of the Insect Nuclear Polyhedroses. Nature **175**, 588—590.
YAMAFUJI, K., 1952: Mechanism of Artifical Virus Formation in Silkworm Tissues. Enzymologia **15**, 223—231.
YARWOOD, C. E., 1952: Latent Period and Generation Time for Two Plant Viruses. Amer. J. Bot. **39**, 613—618.
ZAMECNIK, P. C., E. B. KELLER, J. W. LITTLEFIELD, M. B. HOAGLAND, and R. B. LOFTFIELD, 1956: Mechanism of Incorporation of Labeled Amino Acids into Protein. J. cellul. a. comp. Physiol. (Am.) **47** (Suppl. 1), 81—101.
ZINDER, N. D., 1955: Bacterial Transduction. J. cellul. a. comp. Physiol. (Am.) **45** (Suppl. 2), 23—49.
— and J. LEDERBERG, 1952: Genetic Exchange in Salmonella. J. Bacter. (Am.) **64**, 679—699.

Virus Inclusions in Plant Cells

By

KENNETH M. SMITH

Virus Research Unit
(Agricultural Research Council), Molteno Institute, Cambridge University, Cambridge

With 5 Plates

Contents

Page

Introduction 1
I. The Different Kinds of Plant Virus Inclusions 2
(a) Crystalline or Fibrous Inclusions 2
(1) Intracellular Inclusions 3
(2) Intranuclear Inclusions 3
(b) Amorphous Inclusions (X-bodies) 5
(c) Other Types of Inclusions 6
(d) Distribution of Intracellular Inclusions in Plant Tissues 7
II. Electron Microscopy of Plant Virus Inclusions 9
III. Staining Methods for Intracellular Inclusions 13
References 15

Introduction

IWANOWSKI (1903) was probably the first to point out that certain abnormal intracellular inclusions were always present in some cells of mosaic-infected tobacco plants. We now know that these abnormal inclusions are characteristic of virus infections and do not occur in diseases caused by other infectious agents. Whilst they are not found in all virus diseases, they are invariably present in those in which they do occur. After IWANOWSKI's discovery the next worker to find similar inclusions was probably MATZ (1919) who found plasmodium-like masses in the cells of mosaic-infected sugar cane. In 1921, KUNKEL described intracellular inclusions similar to those described by IWANOWSKI, in maize affected with a mosaic disease. Shortly after this, SMITH (1924) reported the occurrence of amoeboid bodies in the cells of potato plants infected with mild mosaic (potato virus X). Since those days many other workers have described similar inclusions in the cells of plants infected with virus diseases. Among these other early workers may be mentioned the following: GOLDSTEIN (1927) made an extensive study of the inclusion bodies in *Dahlia* plants affected with mosaic and dwarf; HOGGAN (1927) investigated the inclusions in solanaceous plants infected with several different viruses; F. F. SMITH (1926) reported on intracellular inclusions in mosaic tobacco and in the chlorotic areas of mosaic *Datura*. HENDERSON SMITH (1930) made a careful study of the inclusions caused by tomato aucuba mosaic virus in *Solanum nodiflorum*.

Some of these early workers, notably GOLDSTEIN, favoured the theory that the intracellular bodies were some kind of organism, possibly a protozoan, and were the etiological agent of the virus diseases. The majority, however, took the view that the inclusions were an effect rather than the cause of the disease. HENDERSON SMITH (1930) summarized this view—"The alternative is that these bodies are not living organisms but are the reaction-product of the cell to the virus irritant." It is rather interesting in the light of modern knowledge to consider these early speculations and to reflect that those workers who favoured the 'causal organism' theory were more nearly right than those who, more cautiously, preferred the idea of a 'reaction product.'

I. The Different Kinds of Plant Virus Inclusions

(a) Crystalline or Fibrous, (1) Intracellular, (2) Intranuclear.
(b) Amorphous Inclusions (X-bodies).
(c) Other types of inclusions.

(a) Crystalline or Fibrous Inclusions

(1) Intracellular Inclusions

The crystalline inclusions which occur in the cells of plants infected with tobacco mosaic virus are usually in the form of plates which vary considerably in shape. Some are irregular, others may be hexagonal. They are colourless and transparent and have a refractive index higher than the cell sap (BAWDEN 1950). IWANOWSKI showed that the plates develop striations and seem to be composed of needle-like crystals when acidified. He gave them the name 'striate material.' KASSANIS and SHEFFIELD (1941) studied the variations in the kind of cytoplasmic inclusions found by three strains of tobacco mosaic virus. Apart from the crystalline plates and amorphous inclusions previously described, a variety of new forms were induced by all three strains. The new forms were mostly fibrous, one form consisted of spike-like bodies which were needle-shaped with pointed ends and no facets. Another consisted of spindle-shaped bodies made up of aggregates of long fine fibres which might have been composed of a number of the spike-like inclusions of various lengths packed closely side by side. Greatly elongated fibres were also found either alone in a cell or in association with any of the other forms of inclusions. Kassanis and Sheffield consider that all these forms arose either directly or from pre-existing inclusions of the previously recorded types. Variation in the inclusions produced is not apparently due to mutation of the virus. The type of inclusion seems to be determined in a slight degree by the host plant but is largely controlled by the amount of light and heat available to the host.

Besides tobacco mosaic virus and its strains various other viruses give rise to crystalline intracellular inclusions. Isometric crystals, which seem to be similar to the hexagonal crystalline plates of tobacco mosaic virus, have been observed in the cells of *Vicia faba* and three other species of leguminous plants infected with the viruses of pea mosaic and yellow bean mosaic (Plate I *A*). These crystals measure from 0.3 to 0.4 μ in diameter and are plainly isometric; they give the usual reactions for protein and are not found in healthy plants (McWhorter 1941).

Very fine protein crystals have been observed in the cells of virus-infected cacti, *Rhipsalis cereuscula.* These crystals may remain unorientated or grouped into tactoid spindles (Weber, Kenda and Thaler 1952).

(2) Intranuclear Inclusions

This type of inclusion seems to be very rare in the virus diseases of plants although it is common enough in the cell nuclei of insects infected with the nuclear type of polyhedral virus disease.

An interesting case of an intranuclear crystalline inclusion definitely associated with a plant virus is that of severe etch (Kassanis 1939). Both cytoplasmic X-bodies and intranuclear inclusions occur in a number of solanaceous plants infected with severe etch virus. These include *Nicotiana tabacum, N. sylvestris, N. glutinosa, Datura stramonium* and *Hyoscyamus niger* (Plate I *B, C, D*).

The inclusions appear to be produced equally readily in young and old plants but their production is limited to those parts of the leaves showing external symptoms. Most nuclei contain more than one of the crystals and as many as fifteen have been seen in one nucleus. They consist of thin rectangular plates. Their size depends on the size of the nucleus and the number of crystals present, the length varying from 3 to 10 μ. When examined between crossed nicol prisms they show no extinctions when viewed along any axis but it is not certain if this is due to isotropism or because they are too small to give a visible effect.

McWhorter (1941) has observed crystalline inclusions in the nuclei of leguminous plants infected with pea mosaic and yellow bean mosaic. They appear to occur regularly within the nucleoli; the nucleoli of diseased cells assume cubical shapes as they become filled with the isometric crystals. Usually five or more are present but, occasionally, one large crystal fills the entire nucleolus.

Finally Woods and Eck (1948) describe a yellow mottle strain of T. M. V. which gives rise to fibrous and crystalline inclusions in the nuclei of infected cells. The evidence suggests that the nuclear inclusions induced by this strain of T. M. V. are similar or identical to those occurring in the cytoplasm of the same cells. The nuclear and cytoplasmic inclusions induced by this strain reacted in all tests like the various cytoplasmic inclusions found in cells infected with the type virus. It is suggested that this strain of T. M. V. is capable of invading the nuclei of cells.

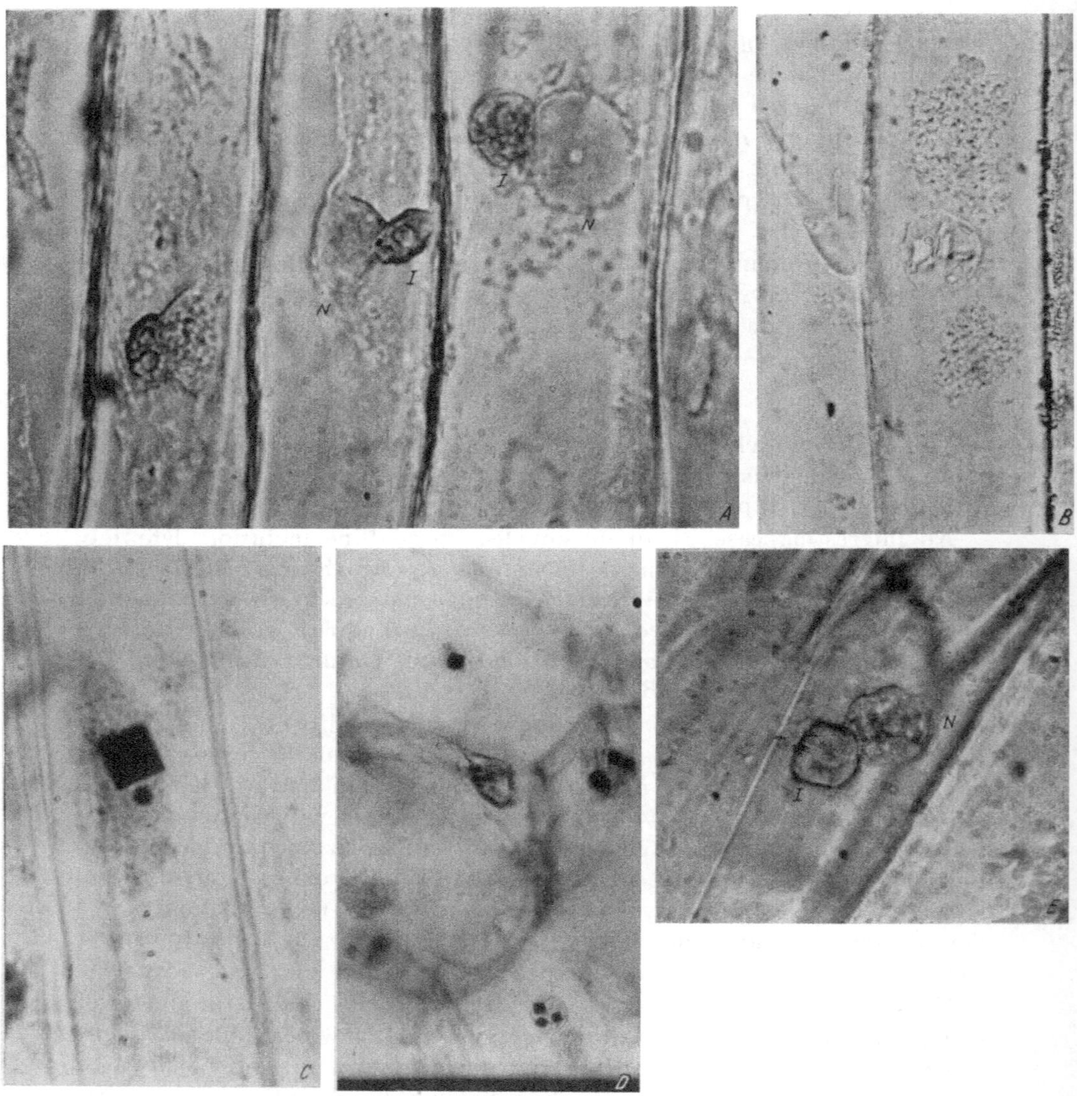

Plate I. *A*, Bean yellow mosaic; *N* = nuclei, *I* = X-body. *B*, Epidermal unstained living cells from tobacco infected with severe etch virus showing two granular X-bodies and a nucleus containing a number of crystalline plates. ×400. *C*, Epidermal cells of tobacco plant, fixed in formol-saline and stained with haematoxylin, showing a crystalline plate. *D*, Pith cells of tobacco plant fixed in formol-saline and stained with haematoxylin, showing crystalline plates in the nuclei; nucleoli unstained. ×400. *E*, Cauliflower mosaic; *N* = nucleus, *I* = X-body.

(*A* after Rubio and van Slogteren; *B*, *C*, *D* after Kassanis, *E* after Rubio.)

(b) Amorphous Inclusions (X-bodies)

Although the amorphous X-bodies are dealt with separately, it is probable that they are very similar in nature at all events in some cases, to the crystalline inclusions previously described. The intracellular inclusions caused by three strains of tobacco mosaic virus have been studied by Bawden and Sheffield (1939). Both types of inclusions, amorphous and crystalline, are produced by these viruses but we are only concerned here with the amorphous X-body. The amorphous bodies of all three diseases are relatively stable and are preserved by ordinary cytological fixatives and give the usual protein reactions. In the case of tobacco mosaic and enation mosaic in which the disease symptoms are similar, the X-bodies are small with a superficial resemblance to amoebae. They average about 10 μ in length and contain vacuoles, chondriosomes and oil globules and as they are carried round in the streaming cytoplasm frequently change their shape. The amorphous body which occurs in the aucuba mosaic disease is larger rather more granular and less like the surrounding cytoplasm. The formation and disappearance of these X-bodies has been examined in detail by Sheffield (1931), who made a cinematograph film of X-body formation. She observed that the rate of streaming of the cytoplasm increased soon after infection of the plant and minute particles of protein appear which are carried about by the cytoplasm of the cell. These particles aggregate and fuse to form large masses which are still carried passively but more slowly about the cell. These fuse until all the protein material is contained in one or sometimes more granular masses. Ultimately the body breaks down giving a number of protein crystals which are presumably similar in nature to the crystalline inclusions previously discussed.

Several other viruses besides that of tobacco mosaic give rise to amorphous intracellular inclusions. Those found in potato plants infected with virus X are similar to the inclusions associated with tobacco mosaic virus (Smith 1924, Clinch 1932).

Plants infected with *Hyoscyamus* mosaic virus contain large numbers of amorphous inclusions in all tissues; these resemble the bodies of aucuba mosaic in appearance and mode of formation. Tobacco ringspot virus also causes the production of large numbers of amorphous inclusions similar to those found in tobacco mosaic (Bawden and Sheffield (1939). Rubio and van Slogteren (1956) describe amorphous intracellular inclusions in the cells of broad bean plants (*Vicia faba*) infected with the virus causing broad bean mottle (Bawden, Chaudhuri and Kassanis 1951). Rubio (1950) has observed intracellular inclusions also in the cells of plants infected with cabbage black ringspot, cauliflower mosaic and turnip yellow mosaic (Plate I *E*). The mode of formation of the bodies appears to vary in the different diseases. In cabbage black ringspot the inclusions develop in a manner similar to that described by Sheffield (1931) for the aucuba mosaic virus. In cauliflower mosaic the inclusions are formed from a single particle by an internal growth and in turnip yellow mosaic from the coalescence in a vacuolated body of degenerate plastids and chloroplasts. RUBIO considers that there is an external membrane round the inclusions of cauliflower mosaic. This question of an external membrane has been raised by several workers, GOLDSTEIN (1924) claimed to have seen them but this was not confirmed by HENDERSON SMITH (1930) and SHEFFIELD (1934).

(c) Other Types of Inclusions

LITTAU and BLACK (1952) have described an unusual type of inclusion body, called a 'spherule,' associated with the wound tumour virus, in the cytoplasm of root tumour cells of sorrel (*Rumex acetosa* L.). These inclusion bodies are also sometimes found in the non-living tumour tracheids but they have never been observed in the non-tumorous cells of roots from which the tumours arose. The spherules are roughly spherical in shape, hyaline and homogeneous. They usually appear to be solid but vacuolated ones do sometimes occur; they range in size from 1 μ up to 4 μ in diameter. In tumour tissue several spherules, varying in size, may occur in a single cell. They have never been observed in normal tissue.

Spherical bodies have also been observed in the split stalks of sugar cane affected with chlorotic streak; they seem to occur in the parenchyma cells surrounding the tracheary vessels (DUFRENOY 1942).

Some nuclear abnormalities have been described by WILKINSON (1953) in the dividing cells of root tips from tomato plants infected with aspermy virus. The nucleolar material instead of dispersing during prophase, persists through anaphase in the form of one or more prominent and somewhat elongated vesicles. WILKINSON puts forward the theory that there is competition between the virus particles and chromonematal material for the nucleoprotein contained in the nucleolus.

There are several instances of spindle-shaped bodies in the cells of plants apparently associated with virus infection, though this association does not seem to have been confirmed in every case. They have been observed in the cells of *Epiphyllum* and *Pereskia* varieties. Grafting and sap-inoculation from plants which showed these inclusions produced similar bodies in plants hitherto spindle-free. The agent which apparently was the cause of these bodies passed a bacterium-proof filter and was destroyed by heating to 100^0 C. Although no external symptoms were apparent in the plants it is thought that a virus is concerned in the production of the spindles (ROSENZOPF 1951).

Somewhat similar "protein spindles" have been observed by WEBER (1953) also in the cells of the cactus *Pereskia aculeata.* These bodies remained intact during the process of autumnal yellowing but underwent drastic shrinkage as soon as the leaves fell and began to decay on the ground.

Examination of leaves of *Opuntia brasiliensis* has revealed numerous fusiform and variably shaped protein bodies in the cytoplasm of the epidermis. Some cells also contain nuclear crystalloids which present many analogies with the protein bodies. All these structures are considered by Miličić (1953) to be virus elements. In a later paper, Miličić and Plavšić (1956) showed that inoculation of a spindle-free *Epiphyllum truncatum* plant with sap of *O. brasiliensis* containing cytoplasmic protein spindles and nuclear crystalloids resulted in the development of both types of inclusions within about three weeks.

In the epidermal cells of the bulb scales of *Lilium tigrinum* protein crystalloids and so-called 'zebra spindles' were observed. These latter appear to be identical with spindles occurring in the Cactaceae and are thought to be of virus origin (Thaler 1956).

Vesicular cell inclusions associated with cabbage black ringspot virus have been described by Rubio (1956). They are spherical or ovoid in form and have a distinct outer membrane. In the early stages, the contents of these inclusions appear to be liquid with scattered refringent particles. Later a crystal-like body is formed and fills most of the space within the membrane. Finally the membrane disappears leaving the crystal-like body intact.

(d) Distribution of Intracellular Inclusions in Plant Tissues

There seems to be a good deal of variation in the type of tissues in which intracellular inclusions occur and this may depend on a number of factors. As a whole, they have been observed in the roots, stems, leaves and flowers of infected plants and in tobacco mosaic they seem to be most abundant in the epidermal and hair cells. In the same disease inclusions have been found regularly in the guard cells of stomata by Hirayama and Yuasa (1935) in Japan, although Sheffield (1936) was unable to find them there. This may be due to differences in climatic conditions since they have now been found also by Kassanis and Sheffield (1941). However, all strains of tobacco mosaic virus do not behave in the same way and inclusion bodies have not been found in the guard cells of plants infected with the aucuba or enation strains.

The species of host plant may also influence the distribution of the inclusions; in *Hyoscyamus niger*, infected with aucuba mosaic, they are confined to the chlorotic areas. In other species they are distributed over the green and yellow tissues. They are very abundant in the hairs, less so in the epidermis and very rare in the palisade and spongy tissues (Sheffield 1931). On the other hand, in the severe etch disease of tobacco inclusions occur in almost every tissue of the plant (Kassanis 1939).

In the case of the broad bean mottle virus, X-bodies were found only in the cells in the chlorotic areas of the leaf of *Vicia faba*, whereas with bean yellow mosaic inclusions occurred in all parts of the leaf and stem of the same host (Rubio and van Slógteren 1956). Here, since the host plant was the same in each case, the distribution of the inclusions seems to depend on the virus concerned.

In a disease known as turnip mosaic caused by a strain of the cabbage black ringspot virus amoeboid bodies abounded but were restricted almost entirely to the subveinal epidermis. Occasionally, one such body was found in cells of the interveinal epidermis (Berkeley and Weintraub 1952).

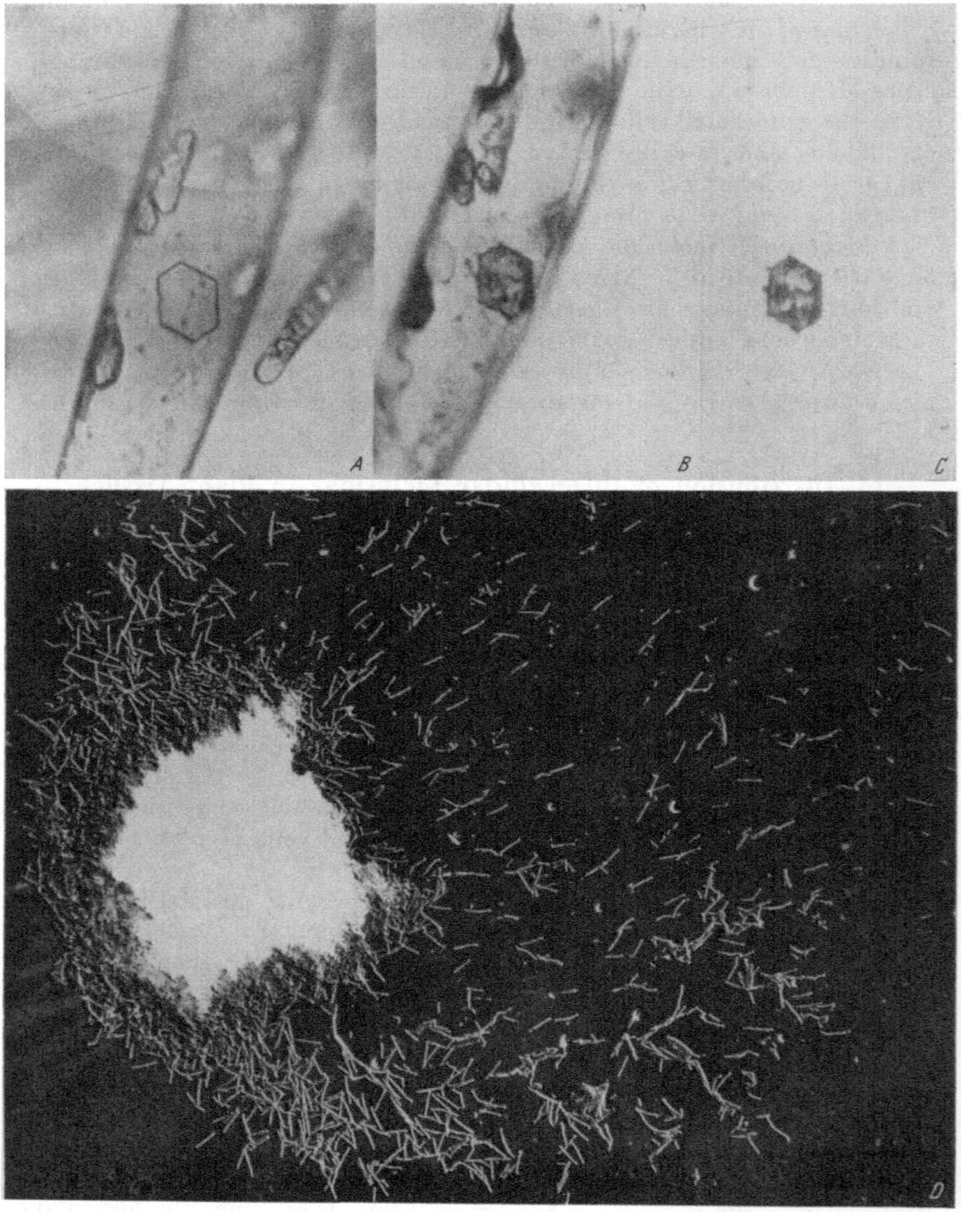

Plate II. *A*, Crystalline inclusion body of tobacco mosaic virus photographed *in vivo* in a hair cell of a Turkish tobacco plant. *B*, The same cell and crystalline body after freeze-drying. *C*, The frozen-dried crystal removed from the cell. ×500. *D*, Electron micrograph of a partially dissolved fragment of hair cell crystal of tobacco mosaic virus showing large opaque residuum of undissolved crystal and numerous well-dispersed particles of tobacco mosaic virus. ×11,000.

(After Steere and Williams.)

II. Electron Microscopy of Plant Virus Inclusion

Beale (1937) made the important suggestion, later to be amply confirmed by electron microscopy, that the crystalline inclusions of T. M. V. were identical with the needle-like crystals obtained by Stanley (1935) in his original purification of this virus. Beale showed that the crystalline plates transformed into needle crystals upon the addition of dilute acid or salt to the water in which epidermal strips had been mounted under the microscope.

Sheffield (1939) attempted to isolate the crystalline inclusions of T. M. V. but was unable to do so because they immediately disintegrated at the touch of the micro-needle. However, this difficulty was overcome by Steere and Williams (1953) who applied the technique of freeze-drying and by this means were able to remove intact the crystalline inclusions. Examination of these inclusions on the electron microscope revealed that the crystals consist apparently of nothing but particles of T. M. V. and a volatile solvent (Plate II *A–D*). Many infective virus particles were present among those comprising the crystalline inclusion bodies. The electron micrographs of the partially dissolved material occasionally showed small areas with considerable regularity of array of the T. M. V. particles. See also Brandes (Plate III *A*).

A further rapid method for the extraction of intact crystalline inclusions from cells of plants infected with T. M. V. has been described (Rubio 1954). This consists in removing strips of epidermis, containing crystalline inclusions, from the leaves and immersing them for 3–4 min. in Carnoy's fluid. They should then be rinsed in 96 per cent. alcohol and placed in distilled water. The intact crystals can then easily be removed by means of two fine needles, and mounted for electron microscopy. If necessary the inclusions can be preserved in 96 per cent. alcohol for a year or more.

Wilkins, Stokes, Seeds and Oster (1950) have reported that the dark bands (striations) appear to interchange position as the crystal is turned through extinction in the polarizing microscope. They conclude that the crystals are built up of thin flat layers lying parallel to the hexagonal face of the crystal, and that within each layer the individual T. M. V. rods are aligned parallel to each other, but with an orientation not quite parallel to that of the T. M. V. rods in the adjacent layers.

Wehrmeyer (1957) has published some excellent electron micrographs of ultra-thin sections illustrating this layer structure (Plate 5, Figs. *A* and *B*).

Steere (1957) has recently perfected a technique which is applicable for detection of some types of structural differentiation within viruses. When applied to the inclusion crystals of T. M. V., it indicates a considerable degree of regularity of orientation of the included virus particles. The particles appear stacked in plates in the manner suggested by Wilkins et al. (1950).

Whilst all the evidence from electron microscopy seems to suggest that

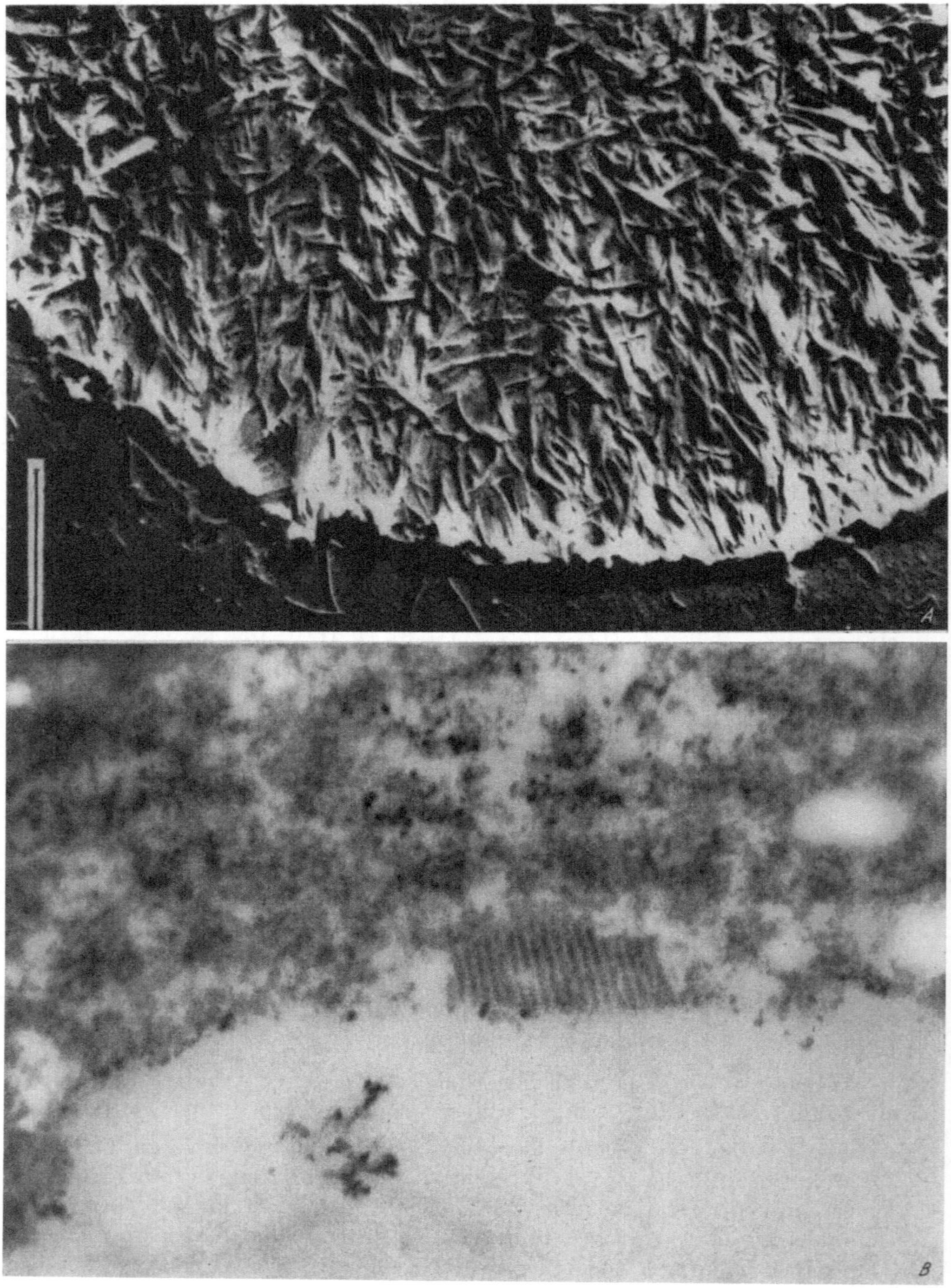

Plate III. *A*, X-body of tobacco mosaic virus showing the virus rods. ×25,000. *B*, Thin section of leaf palisade cell of *Datura stramonium* infected with tomato bushy stunt virus; note the apparent crystalline arrangement of the virus particles. ×55,500.

(*A* after Brandes; *B* after K. M. Smith.)

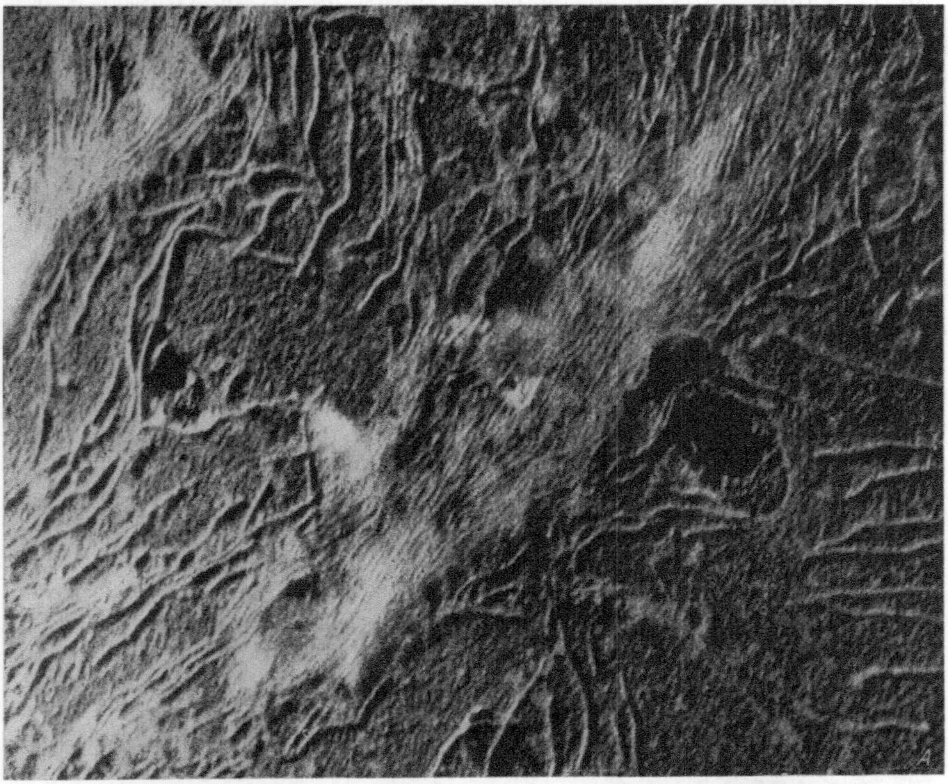

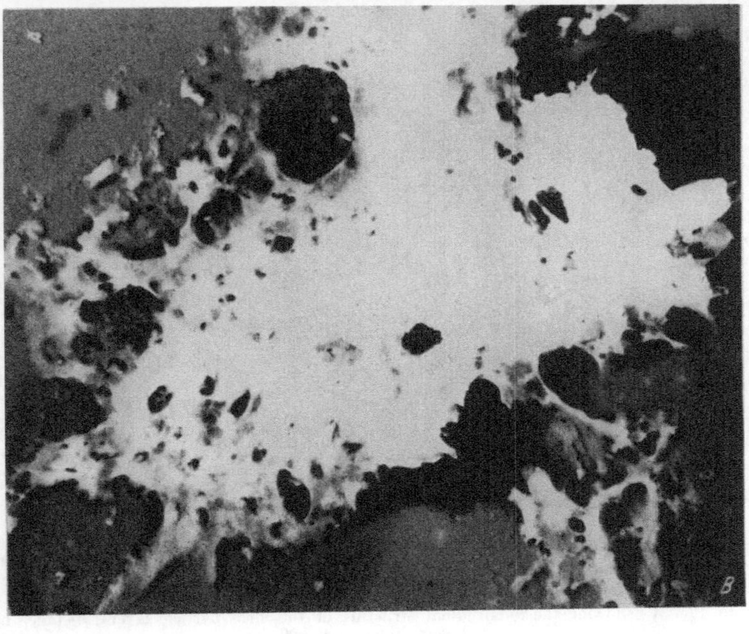

Plate IV. *A*, Cabbage black ring; fragment of inclusion body formed by long flexuous particles of virus. *B*, X-body from bean yellow mosaic on *Vicia faba*, no virus particles can be detected.
(*A* and *B* after RUBIO.)

the crystalline intracellular inclusions consist mainly of virus, it is less clear cut as regards the amorphous X-bodies. In the case of the broad bean mottle virus, electron micrographs (Rubio and van Slogteren 1956) show the X-bodies to be composed mainly or entirely of spherical particles

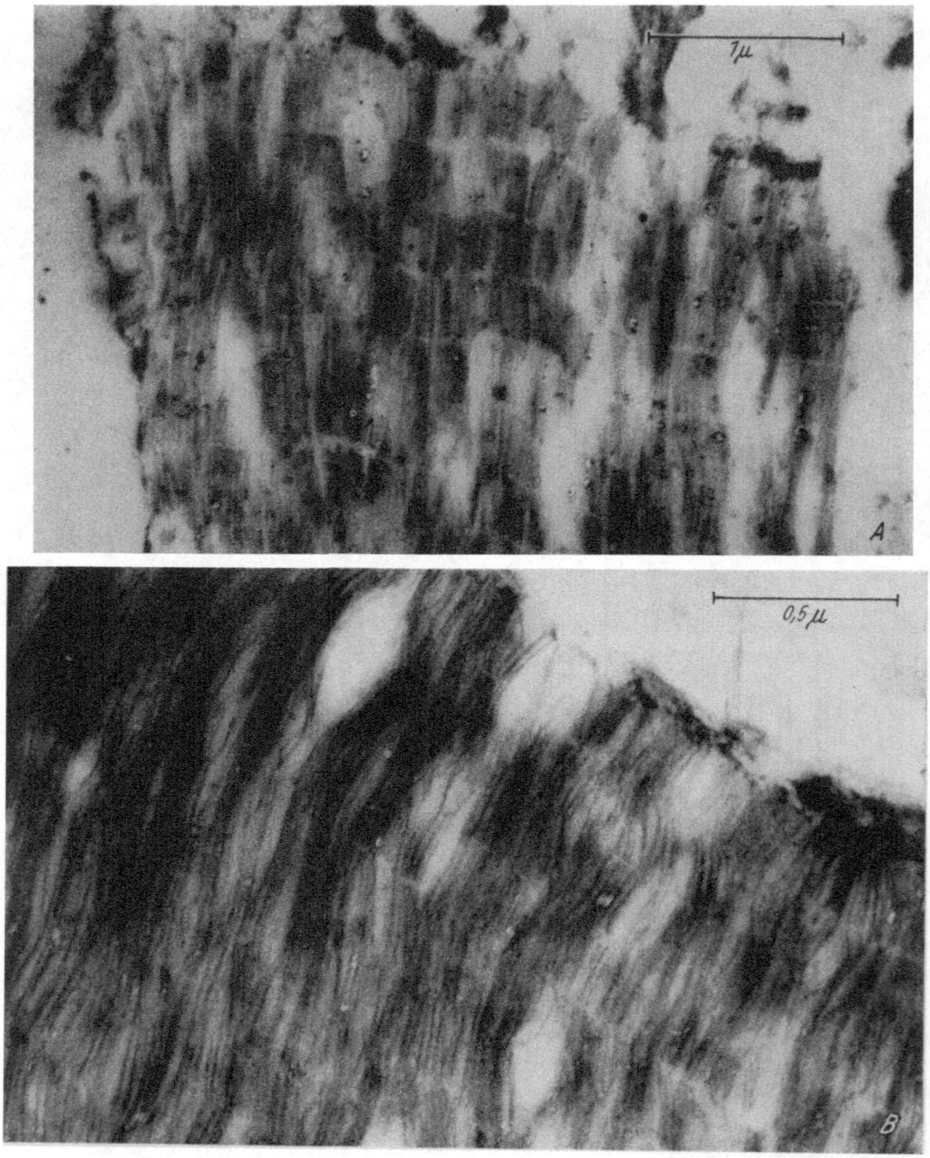

Plate V. *A*, Ultrathin section of an hexagonal prism of tobacco mosaic virus (T. M. V.), longitudinal view showing the layer structure; length of virus particles about 300 mμ. ×26,100.
(After Wehrmeyer, 1957.)

B, Similar to Fig. *A*, showing the longitudinal structure of the virus particle layers; distance between the layers 20 mμ. ×49,000.
(After Wehrmeyer, 1957.)

which, both in size and shape, are apparently identical with the broad bean mottle virus in purified preparations (BAWDEN, CHAUDHURI and KASSANIS 1951). Similarly, X-bodies in a late stage of the cabbage black ring disease consisted mainly of flexuous strands resembling virus particles (Plate IV *A*). Again, the intracellular inclusions of henbane mosaic appear to be composed of rod-shaped virus particles. On the other hand, the X-bodies of cauliflower mosaic and of a common strain of T. M. V. were found to contain few if any detectable virus particles and appeared to consist mainly of amorphous material (RUBIO 1950, 1956).

In examining by electron microscopy ultra-thin sections of virus-infected plant cells, it is usually very difficult to differentiate the smaller near-spherical viruses from the normal cytoplasm of the cells. However, special circumstances may arise which would make identification of the virus particles possible. Such a special circumstance would be the aggregation of the virus or the formation of some kind of intracellular inclusion. Examination of virus-infected cells of *Datura stramonium* under the electron microscope has been made by SMITH (1956), the virus in question being that of tomato bushy stunt. On one or two occasions groups of particles in a regular alignment were observed on the inner edge of the cytoplasm of certain cells, usually palisade cells. This alignment is obviously a crystalline arrangement and the spacing, 30 mμ, is the same as the spacing in the crystals of tomato bushy stunt virus (Plate III *B*). It is suggested therefore that these groups of particles, which so far have never been observed in healthy plants are in fact minute virus crystals presumably similar in character to the much larger crystalline intracellular inclusions.

III. Staining Methods for Intracellular Inclusions

A number of workers notably BALD (1949 a) have used fixatives and staining techniques for the study of intracellular inclusions. BALD found the following fixatives would preserve the inclusions in the cells of infected plants and prepare them to accept appropriate stains, Lugol's iodine-potassium iodide, the same in 50 per cent. alcohol (either of which may be combined with 5 per cent. formalin or with chromic acid), and a third series, the effect of which depends on the action of concentrated acetic acid and cerium nitrate on the virus. Some of these fixatives preserve the structure of plastids, mitochondria and other cell constituents that may be important in virus multiplication. In addition to Giemsa stain, iron alum haematoxylin, various acid fuchsin combinations, Flemmings triple stain, Nile blue sulphate and trypan blue were used.

Inclusion bodies and more disperse material, presumed to be virus, were stained uniformly. Tissues containing tobacco mosaic virus, potato virus X, a mixture of each, the viruses of tomato spotted wilt, squash mosaic, potato leaf-roll, and alfalfa dwarf disease all gave similar staining reactions. BALD (1949 b) has also fixed sections of tobacco leaves diseased with

T. M.V. in Karpechenko's fluid, pretreated them with iodine-potassium iodide solution and stained with Giemsa-Orange G. The amorphous inclusion bodies, as well as other virus inclusions, stain purple and the host nucleus stains greenish blue to purple.

McWHORTER (1940) has used the stain trypan blue to identify virus inclusions. The tissues of tulips infected with tulip mosaic virus contain intracellular inclusions which stain characteristically with trypan blue. On the other hand, lily latent virus also infects tulip producing almost identical leaf symptoms and abundant X-bodies in the cells. The latter, however, do not stain with trypan blue. This is a possible method of distinguishing between two viruses causing similar symptoms in the same host plant.

Trypan blue and phloxine have been used with success by BERKELEY and WEINTRAUB (1952) for staining inclusion bodies in the epidermal cells of turnips and *Nicotiana glutinosa* infected with turnip mosaic (cabbage black ringspot virus). Zenkers Fluid and 10 per cent. formalin have been used for fixing tissues of stock, *Matthiola incana,* infected with what is probably also cabbage black ringspot virus. Satisfactory staining was achieved with the use of phloxine or trypan blue or both, but for photomicrography Altmann's acid fuchsin is recommended (SARDINA, CORDON and HUERTOS 1949).

Staining and cytochemical reactions of the spherules in sorrel virus tumours have been investigated by LITTAU and BLACK (1952). The spherules stained red with safranin after treatment with Flemming's triple stain or safranin and fast green. After immersion in solutions of pyronin and methyl green, the bodies were stained red by the pyronin. They stained red with acid fuchsin but were not stained by trypan blue. In all these staining reactions, the spherules reacted just as did the nucleoli. The spherules did not stain with iodine potassium iodide solution nor with Sudan IV. Their staining reaction with pyronin was unchanged after extraction in ether for 6 hours. They were not stained by the Feulgen reaction. Differential staining of the spherules and the nucleoli was achieved by the use of azure A which stained the nucleoli slightly more intensely than the spherules. Clear differentiation was attained by a nitrous acid-ribonuclease-azure A technique which stained the spherules light blue and the nucleoli dark blue.

This last procedure was the only satisfactory method of distinguishing spherules from nucleoli except for the relative positions, in the cytoplasm and nuclei respectively.

Flemming's triple stain, safranin and fast green, or pyronin and methyl green, were excellent mixtures for revealing the presence of the spherules.

For the examination of ultrathin sections of the crystalline inclusions of tobacco mosaic, WEHRMEYER (1957) fixed the material in osmic acid and obtained a contrast effect by means of uranyl acetate according to the method of STRUGGER (1956).

References

AMELUNXEN, F., 1956: Über die Strukturanalyse der Eiweißspindeln der Cactaceen. Protoplasma **45**, 228.

— 1958: Die Virus-Eiweißspindeln der Kakteen. Darstellung, elektronenmikroskopische und biochemische Analyse des Virus. Protoplasma **49**, 140.

BALD, J. G., 1949 a: Additional methods for fixing and staining viruses in infected plant tissues. Amer. J. Bot. **36**, 335—342.

— 1949 b: A method for the selective staining of viruses in infected plant tissues. Phytopathology **39**, 395—402.

BAWDEN, F. C., 1950: Plant Viruses and Virus Diseases. 3rd ed. Chronica Botanica Co., U. S. A.

— R. P. CHAUDHURI, and B. KASSANIS, 1951: Some properties of broad-bean mottle virus. Ann. appl. Biol. **38**, 774—784.

— and F. M. L. SHEFFIELD, 1939: The intracellular inclusions of some plant virus diseases. Ann. appl. Biol. **26**, 102—115.

BEALE, H. PURDY, 1937: Relation of Stanley's crystalline tobacco virus protein to intracellular crystalline deposits. Contr. Boyce Thompson Inst. Pl. Res. **8**, 415—431.

BERKELEY, G. H., and M. WEINTRAUB, 1952: Turnip mosaic. Phytopathology **42**, 258—260.

DUFRENOY, J., 1942: Vacuolar inclusions in cells of sugar cane affected with chlorotic streak. Phytopathology **32**, 3.

GOLDSTEIN, B., 1924: Cytological study of living cells of tobacco plants affected with mosaic disease. Bull. Torrey Bot. Club **51**, 261—273.

— 1927: The X-bodies in the cells of *Dahlia* plants affected with mosaic disease and dwarf. Bull. Torrey Bot. Club **54**, 285—295.

HIRAYAMA, S., and A. YASA (1935): Cytological study of tobacco mosaic I. Ann. Phytopath. Soc. Japan **5**, 197—205.

HOGGAN, I. A., 1927: Cytological studies on virus diseases of solanaceous plants. J. Agric. Res. **35**, 651—672.

IWANOWSKI, D. J. V. V., 1903: Über die Mosaikkrankheit der Tabakpflanze. Z. Pflanzenkrkh. **13**, 1—14.

KASSANIS, B., 1939: Intranuclear inclusions in virus infected plants. Ann. appl. Biol. **26**, 705—709.

— and F. M. L. SHEFFIELD, 1941: Variations in the cytoplasmic inclusions induced by three strains of tobacco mosaic virus. Ann. appl. Biol. **28**, 360—367.

KUNKEL, L. O., 1921: A possible causative agent for the mosaic disease of corn. Hawaiian Sug. Plant. Assoc. exper. Sta. Bot. Ser. Bul. **3**, 1—14.

LITTAU, C. VIRGINIA, and L. M. BLACK, 1952: Spherical inclusions in plant tumors caused by a virus. Amer. J. Bot. **39**, 87—95.

MATZ, J., 1919: Infection and nature of the yellow stripe disease of cane. J. Dept. Agric. Puerto Rico **3**, 65—82.

MCWHORTER, F. P., 1940: Separation of Tulip Virus 1 from lily latent virus by cytological methods. Phytopathology **30**, 788.

— 1941: Isometric crystals produced by *Pisum Virus* 2 and *Phaseolus Virus* 2. Phytopathology **31**, 760—761.

MILIČIĆ, D., 1953: Virus bodies and cell division anomalies in *Opuntia brasiliensis* (Transl.). Protoplasma **42**, 228—236.

— 1956: Virus-Zelleinschlüsse in *Alliaria officinalis*. Protoplasma **47**, 341—346.

— and B. PLAVŠIĆ, 1956: Eiweißkristalloide in Kakteen-Virusträgern. Protoplasma **46**, 547—555.

REITER, L., 1957: Vitale Fluorochromierung pflanzlicher Viruseinschlußkörper. Protoplasma **48**, 279.

ROSENZOPF, E., 1951: Sind Eiweißspindeln Virus-Einschlußkörper? Ber. ges. Biol. A. **75**, 3—4.

RUBIO, H., 1950: Estudios sobre inclusiones intracelulares producidas por virus en las plantas. Microbiol. Española **3**, 207—232.

— 1954: Rapid extraction of intact crystalline inclusions from the cells of plants infected with tobacco mosaic virus. Nature Lond. **174**, 313.

— 1956: Origin and composition of cell inclusions associated with certain tobacco and crucifer viruses. Phytopathology **46**, 553—556.

Rubio, H., and D. H. M. van Slogteren, 1956: Light and electron microscopy of X-bodies associated with broad-bean mottle virus and *Phaseolus Virus* 2. Phytopathology **46**, 401—402.

Sardina, R., C. F. Martinez, and H. Rubio, 1949: Consideraciones acerca de las tecnicas de tinción de inclusiones en las virosis vegetales. Bol. Pat. veg. Ent. agric. Madrid **16**, 311—320.

Sheffield, F. M. L., 1931: The formation of intracellular inclusions in solanaceous hosts infected with aucuba mosaic of tomato. Ann. appl. Biol. **18**, 471—493.

— 1934: Experiments bearing on the nature of intracellular inclusions in plant virus diseases. Ann. appl. Biol. **21**, 430—453.

— 1936: The susceptibility of the plant cell to virus disease. Ann. appl. Biol. **23**, 498—508.

— 1939: Micrurgical studies on virus-infected plants. Proc. r. Soc. Lond. B. **126**, 529—538.

Smith F. F., 1926: Some cytological and physiological studies of mosaic diseases and leaf variegations. Ann. Miss. Bot. Gard. **13**, 425—484.

Smith, J., Henderson, 1930: Intracellular inclusions in mosaic of *Solanum nodiflorum*. Ann. appl. Biol. **17**, 213—222.

Smith, Kenneth M., 1924: On a curious effect of mosaic disease upon the cells of the potato leaf. Ann. Bot. **38**, 385.

— 1956: Apparent crystallization of tomato bushy stunt virus in the cells of infected plants. Virology **2**, 706—709.

Stanley, W. M., 1935: Isolation of a crystalline protein possessing the properties of tobacco virus. Science **81**, 644—645.

Steere R. L., 1957: In the Nature of Viruses. Ciba Found. Symp. J. & A. Churchill, p. 24.

— and R. C. Williams, 1953: Identification of crystalline inclusion bodies extracted intact from plant cells infected with tobacco mosaic virus. Amer. J. Bot. **40**, 81—84.

Strugger, S., 1956: Die Uranylacetat-Kontrastierung für die elektronenmikroskopische Untersuchung von Pflanzenzellen. Naturwiss. **43**, 357.

Thaler, I., 1956: Proteinspindeln und anormale Zellwandbildung in der Epidermis viruskranker *Impatiens holstii*-Pflanzen. Protoplasma **46**, 755—761.

Weber, F., 1953: Eiweißpolyeder in *Pereskiopsis*-Virusträgern. Protoplasma **42**, 283—286.

— G. Kenda und I. Thaler, 1952: Viruskörper in Kakteen-Zellen. Protoplasma **41**, 277—286.

Wehrmeyer, W., 1957: Darstellung und Strukturordnung eines Tabakmosaikvirus-Einschlußkörpers in der Zelle. Naturwiss. **44**, 519.

Wilkins, M. H. F., A. R. Stokes, W. E. Seeds, and G. Oster, 1950: Tobacco mosaic virus crystals and three-dimensional microscopic vision. Nature, Lond. **166**, 127—129.

Wilkinson, J., 1953: Virus-induced nucleolar abnormalities in tomato. Nature, Lond. **171**, 658.

Woods M. W., and R. V. Eck, 1948: Nuclear inclusions produced by a strain of tobacco mosaic virus. Phytopathology **38**, 852—856.

Zech, H., 1952: Infektionsvorgang und Wanderung des Tabakmosaikvirus. Planta **40**, 461—514.

Protoplasmatologia
IV. Virus
4. Virus-Einschlußkörper
b) Virus Inclusions in Insect Cells

Virus Inclusions in Insect Cells

By

KENNETH M. SMITH

Virus Research Unit
(Agricultural Research Council), Molteno Institute, Cambridge University, Cambridge

With 16 Figures

Contents

	Page
Introduction	1
I. Polyhedral Virus Diseases	2
Polyhedral Diseases of Lepidoptera	3
Nuclear Polyhedra	4
Cytoplasmic Polyhedra	8
Double Virus Infections	10
Virus Infection of Pupae and Adults	10
Polyhedral Disease of a Dipterous Larva	11
Polyhedral Diseases of Hymenoptera	14
II. Granuloses or Capsular Diseases	18
III. Viruses without Polyhedral or Granular Inclusions	19
References	24

Introduction

Although the virus diseases of insects have been known for many years, investigation of them has been rather spasmodic and it was not until the development of the electron microscope that our knowledge of these viruses began to expand. There are several distinct kinds of viruses affecting insects, and with further investigations, no doubt others will be discovered.

The largest and most-studied group of viruses are those causing the polyhedral diseases or polyhedroses. These diseases are characterized by the presence in the blood and certain tissues of large numbers of many-sided crystals, or *polyhedra*. A second large group of viruses causes the diseases known as granuloses or capsular diseases in which numbers of very small *granules* are produced in the affected larvae.

Finally, there are the viruses which are not enclosed in crystals or granules but may nevertheless give rise to a kind of intracellular inclusion. A remarkable virus of this type affecting the dipterous larva, *Tipula paludosa,* is described in some detail.

For many years the polyhedral crystals were the subject of much controversy and discussion. They were first considered to be some kind of infectious agent or "chlamydozoan," then to be carriers of the infective agent, KOMAREK and BREINDL (1924) and others, and finally a crystalline accumulation of the virus (PAILLOT and GRATIA 1939, BERGOLD 1943).

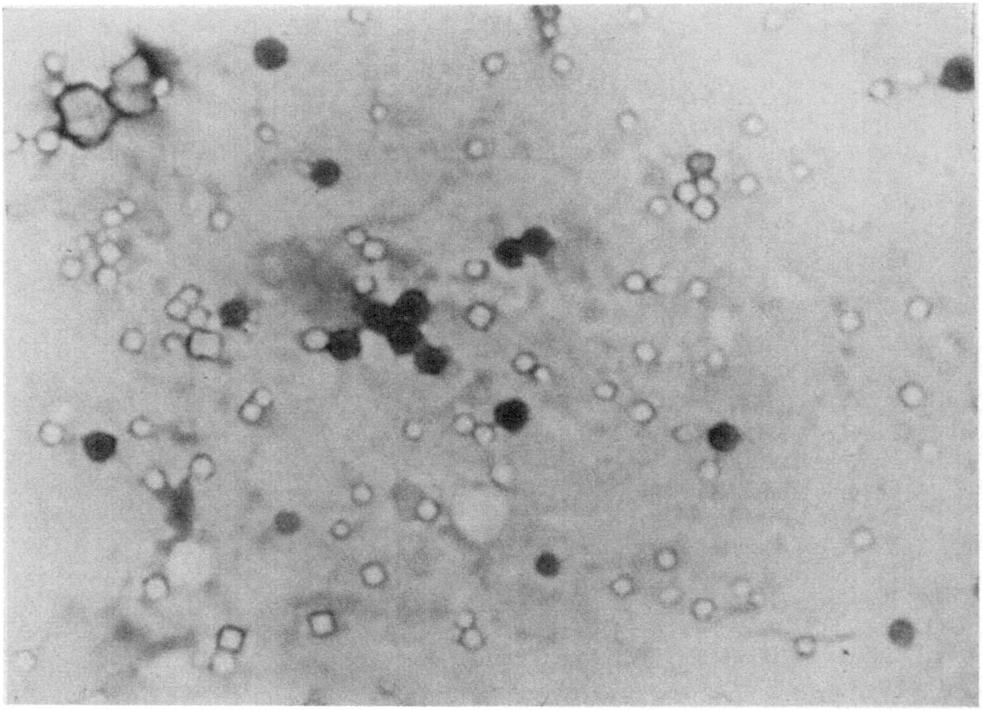

Fig. 1. Blood smear from a larva of *Sphinx ligustri* infected with both a nuclear and a cytoplasmic polyhedrosis; note the cytoplasmic polyhedra are stained. ×850.

It is now known that the virus particles are contained within the polyhedral crystal as suggested by KOMAREK and BREINDL (1924) using the optical microscope and demonstrated by BERGOLD (1947) on the electron microscope.

I. Polyhedral Virus Diseases

Like the majority of insect viruses, the polyhedral viruses attack the larval forms, although cases of infection of adult moths have been recorded. These viruses are known to attack the larvae of the Lepidoptera, Hymenoptera and Diptera, but infections of Lepidoptera are by far the most numerous. In this study it is proposed to take the different groups of viruses in turn, beginning with the polyhedral viruses.

Polyhedral Diseases of Lepidoptera

The polyhedral diseases are often referred to as 'nuclear diseases,' because it was thought that the polyhedra always began to develop in the cell nucleus of certain tissues. It has been shown recently, however, (SMITH and WYCKOFF 1951, XEROS 1952, SMITH and XEROS 1953 a) that this is

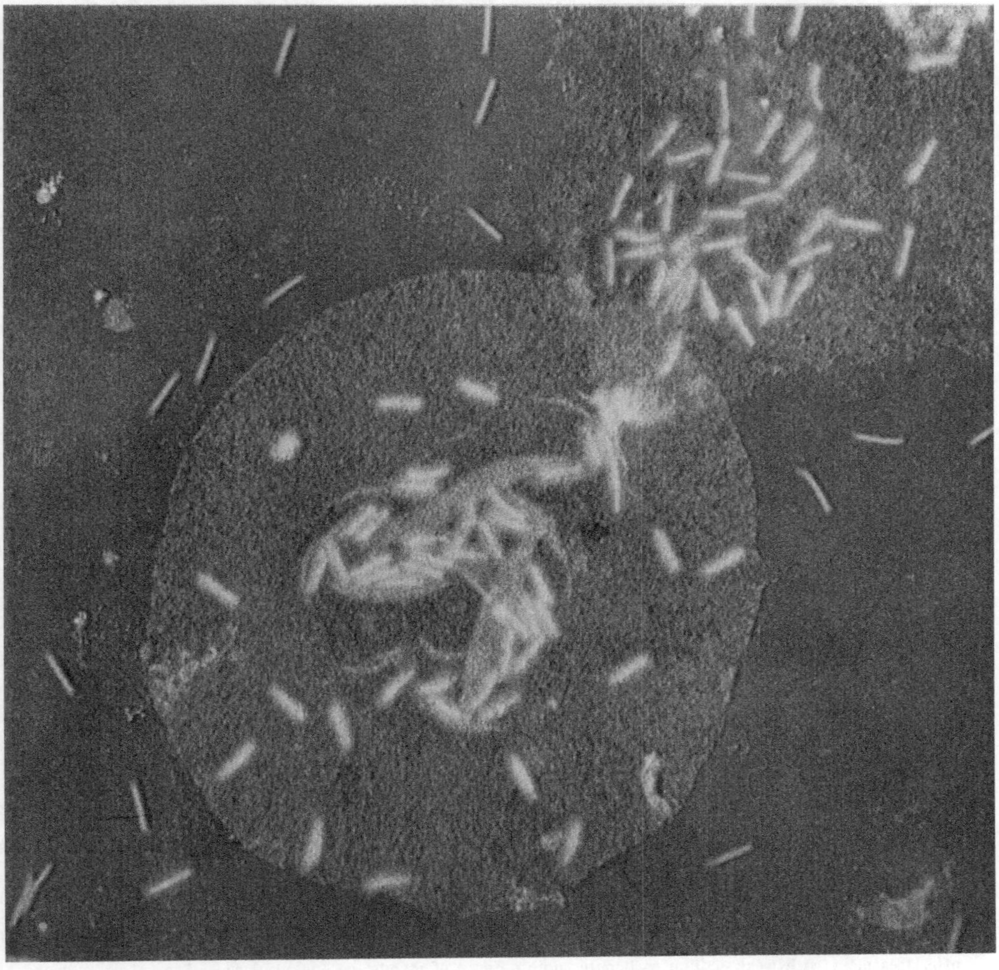

Fig. 2. Nuclear polyhedral crystal from an infected noctuid larva after treatment with weak sodium carbonate; note the membrane which enclosed the crystal, now dissolved. the virus rods each inside a capsule and, outside, the virus rods free of the enclosing capsule. ×18,300.

not the case, and that there exist at least two entirely separate types of polyhedral viruses attacking lepidopterous larvae. These two types cause the *nuclear* and *cytoplasmic* polyhedroses; the first named is the classical disease which has been extensively studied in the silkworm, *Bombyx mori* (L.), the nun moth, *Lymantria monacha* (L.) and the gypsy moth, *Porthetria dispar* (L.). The cytoplasmic viruses, though only recently recognized, are extremely common and have been described in a large number of caterpillars of both moths and butterflies. Each of these two types of viruses has its characteristic disease symptoms, polyhedral body and location in the tissues (Figs. 2 and 3).

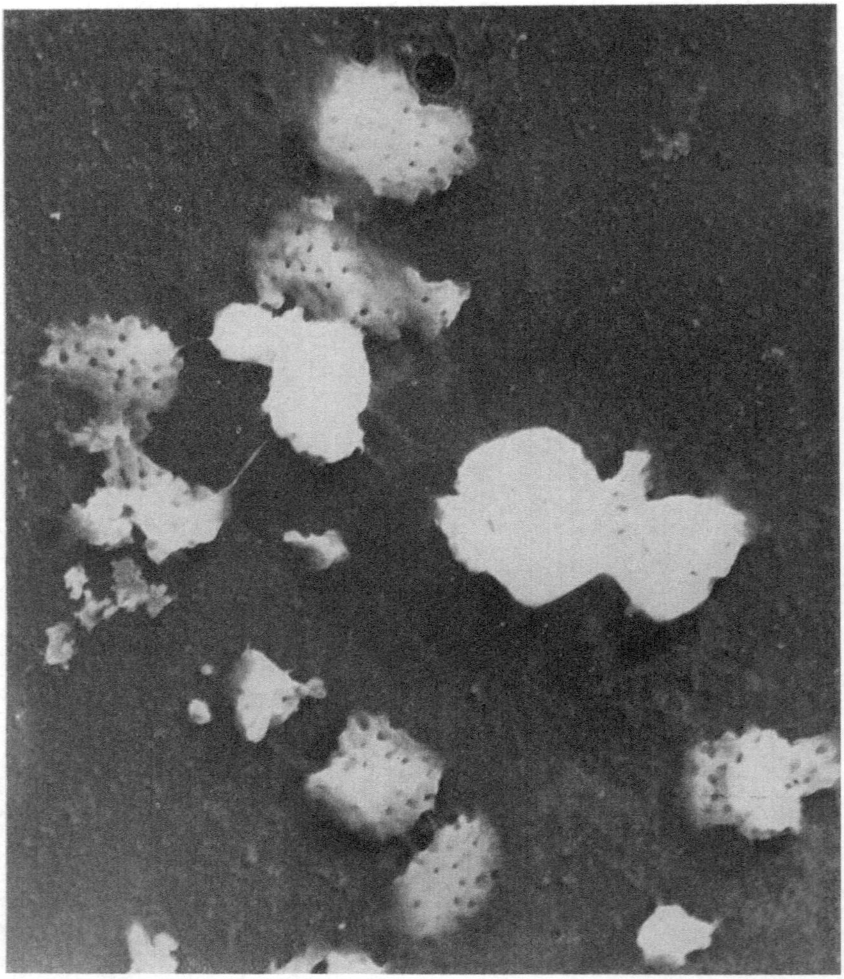

Fig. 3. Cytoplasmic polyhedral crystal from infected larva of *Phlogophora meticulosa*, the angleshades moth, after treatment with weak sodium carbonate; note absence of membrane and virus rods. The spherical virus bodies are thought to have been contained in the round holes which can been seen. (Compare Figs. 10 and 11.) ×27,000.

Nuclear Polyhedra

(a) Morphology. The polyhedra vary greatly in size and shape, and may be many sided, rectangular or cubic. The size depends on the stage of growth of the polyhedral crystal and may vary from the limit of optical vision to 15 or more microns in diameter. It is not yet known for certain how far the shape of the polyhedral crystal is characteristic of the virus or the host species. But it has been observed that in certain species, the scarlet tiger moth *Panaxia dominula,* for example, the polyhedra are always rectangular. It has also happened in experiments that the polyhedra developing in the inoculated insect have been characteristic of those from the source insect. It is usual for all the polyhedra in an individual nucleus to be of the same size, though the size varies in different nuclei. This is because the polyhedra are apparently all formed at once and subsequently only increase in size. There seems to be no continuous pro-

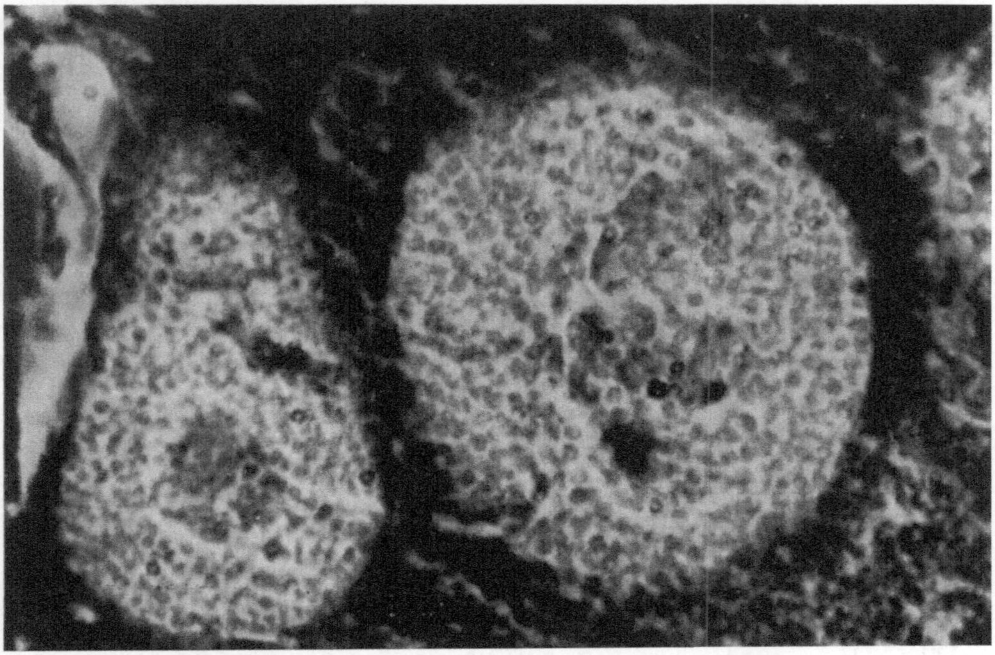

Fig. 4. Section through the silk glands of the larva of *Tinea bisselliella,* the clothes moth, infected with a nuclear polyhedrosis; note the polyhedra inside the silk glands. ×1,750. (SMITH and XEROS)

duction of new polyhedra in a given nucleus and the number is clearly correlated with the size of the nucleus (SMITH and XEROS 1953 b).

(b) Location of Nuclear Polyhedra in the Tissues. In the nuclear diseases the polyhedra are usually confined to the cells of the epidermis, tracheae and fat-body, and do not as a rule occur in the gut. This, as we shall see later, is very different from the cytoplasmic polyhedra. However, there are exceptions to this rule and these are well shown by the nuclear polyhedroses of two species of clothes moth, *Tinea bisselliella* and *Tineola pellionella.* Sections of these larvae infected with a nuclear polyhedrosis revealed the presence of polyhedra in the nuclei of cells of the gut, the silk glands (Fig. 4), the gonads and wing buds (SMITH and XEROS 1954 a). The presence of polyhedra in the cytoplasm of gut cells of *T. bisselliella* had previously been pointed out by LOTMAR (1941) but since they were in the cytoplasm they were most probably due to the second type of polyhedral virus which had not at that date been recognized.

(c) Nature of the Nuclear Polyhedra. The nuclear polyhedra are protein in nature, but are not nucleo-proteins, and are apparently serologically unrelated to the virus. BERGOLD (1953) has suggested three possible alternatives to explain the nature of the polyhedral inclusions: —(1) The polyhedral protein is a part of the virus, not fully transformed due to lack of nucleic acid. (2) The polyhedral protein is a metabolic by-product of the virus. (3) The polyhedral protein is a metabolic product of the host, as first suggested by VON PROWAZEK (1907). BERGOLD considers that the first proposition is refuted by the differences in amino acid composition in the virus and polyhedra respectively and quotes WELLINGTON (1951). He considers the second alternative unlikely in view of the large quantity of the polyhedral protein in comparison with the virus itself. The third possibility is supported by the fact that the formation of the polyhedral bodies does immobilize, and prevent the development of the virus.

To these suggestions the following comments may be added; in sections of infected nuclei of *B. mori,* the silkworm, the polyhedral protein can be seen forming and depositing round the virus rods in a manner which seems quite independent of the virus itself (SMITH and XEROS 1954 a). If the production of the polyhedral bodies is a defensive mechanism on the part of the host it is an extremely ineffective one. Once the polyhedral bodies are formed the caterpillar invariably dies, and, moreover, the polyhedra themselves act as a most efficient mechanism for the spread of the virus.

For a more detailed account of the chemical nature of the polyhedra, the reader is referred to BERGOLD (1953).

In a later section dealing with the cytoplasmic polyhedra, it is shown that their staining properties offer a ready means of differentiation from the nuclear polyhedra, which do not stain with Giemsa solution or methylene blue. However, by means of a modification of the cytochemical methods used by MAZIA et al. (1953), XEROS has shown that intense staining of the nuclear polyhedra is achieved on Carnoy-fixed sections by pretreatment with 1 N hydrochloric acid for ten minutes at 60° C. With bromophenol blue, the staining properties of the nuclear polyhedra at all stages are identical after a great variety of pretreatments of the sections. The presence of mercury in the bromophenol blue is not necessary for intense staining.

By means of special staining methods it is possible to show up optically the virus rods or bundles of rods in certain cases inside the polyhedra. For example, sections of larvae of *P. meticulosa,* the angleshades moth, infected with a nuclear polyhedrosis, were stained with Giemsa solution overnight, following treatment with 3 per cent. nitric acid in 93 per cent. alcohol for one day at room temperature. After this treatment the polyhedra stained a pink colour, and the virus bundles inside them were distinctly visible; they were purple in colour and varied in size from 0.2 μ to 0.5 μ or less in diameter (SMITH and XEROS 1953 b).

The reactions of polyhedra from different species to the action of alkalis, such as weak sodium carbonate, vary greatly. Some are dissolved by ½ per cent. sodium carbonate for 1 min., others by 5 min. treatment with 4 per cent. sodium carbonate, whilst the nuclear polyhedra of *Pterolocera amplicornis* withstand this treatment for 30 min., and an exposure of at least 60 min. at 56° C. is necessary to dissolve them (DAY et al. 1953).

(d) Formation of Virus in the Nucleus. The nuclear polyhedroses of insects offer a unique opportunity for the study of the development of the virus and its absorption into the polyhedral crystal. Since the nucleus is known to be the chief focus of multiplication of the virus, attention can be directed to a small and very definite area. This can best be done by the examination, on the electron microscope, of very thin sections of diseased nuclei in varying stages of infection (SMITH and XEROS 1953 c, 1954 a; HUGHES 1953).

In diseased nuclei of the silkworm *Bombyx mori,* infected with a nuclear polyhedrosis, there arises at a certain stage of the disease a central chromatic mass or "nuclear net." Previous to this the nucleus becomes greatly enlarged. Next, the chromatic bodies begin to clump together to form the nuclear net; at this stage no virus rods are visible. A little later, the rods begin to form and can be seen protruding from the central chromatic net and from the peripheral chromatin. Sections give the appearance of several bundles of virus rods being protruded from the central chromatic mass and from peripheral chromatin bodies into the ring zone where the individual rods separate from one another. Sometimes a long virus body about 500 mμ in length can be seen protruding from the peripheral chromatin. Similar rods about twice the normal length of virus rods have also been observed; in some cases they give the appearance of breaking across to form smaller sized rods.

The distribution of the virus rods in the infected nuclei shows several features which suggest that the virus rods form in the central and peripheral chromatic material, largely in the former. A great proportion of the virus rods are found partially embedded in the chromatic material, presumably in the process of slipping out of these masses. In very thin sections at high magnification it can be seen that the chromatic mass has a very marked, fine fibrillar structure in addition to a smoothly deposited material. The fibrils of the chromatic mass are found to be of varying thickness up to the diameter of the virus rods, from which some of the fibrils seem indistinguishable. The evidence suggests that the virus rods may arise in the chromatic masses by differentiation into fibrils of anomalous pre-virus materials which accumulate in the chromatic masses. The possibility that long fibrils may also differentiate and later break down to normal or smaller sized virus rods is also suggested. It seems that the virus rods are then freed or protruded into the ring zone where they accumulate.

After the appearance of free virus rods in the nuclear ring zone, the polyhedral protein begins to accumulate in the ring zone and to aggregate around the virus rods. In some of the sections, it is possible to see the rods actually being embedded in the polyhedral protein, giving a sculptured effect.

In the final stage the nucleus becomes filled with completed polyhedra, the nuclear sap being again free of polyhedral protein and virus rods.

A similar study on the development of a nuclear virus in insects has been published by Hughes (1953). In this case the disease was a polyhedrosis of alfalfa caterpillar, *Colias philodice enrytheme* Bdvl. Rod-shaped virus particles appear in large numbers in the nucleus of an infected cell. The rods form bundles of two or more members, apparently surrounded by a membrane. A dense substance appearing within the membrane renders that structure opaque. Deposition of polyhedral protein around and between a number of bundles results in the formation of a small polyhedron containing bundles of virus particles. The polyhedron apparently grows by the progressive deposition of polyhedral protein at its periphery and the entrapping of adjacent bundles within its mass.

Consecutive serial sections of polyhedra obtained from *Porthetria dispar* virus revealed bundles of rods scattered at random within the polyhedral body. Each bundle was entirely surrounded by a dense, sharply defined membrane. The rods measured 18–22 mμ in diameter and averaged 280 mμ in length (Morgan, Bergold, and Rose 1956).

Cytoplasmic Polyhedra

(a) Morphology of Polyhedra. In studies upon the polyhedral diseases of the larvae of two species of tiger moths, *Arctia villica* and *A. caja,* (Smith and Wykoff 1951) an entirely different kind of polyhedral body from those previously known was described. Later these were shown to be associated with a different type of symptom and to develop in the cytoplasm and not in the nucleus (Xeros 1952, Smith and Xeros 1953 a).

Superficially on the optical microscope the cytoplasmic polyhedra seem very similar to the nuclear type. On the whole they tend to be larger than the latter, and may sometimes be almost spherical in shape. It has been found that, unlike the nuclear polyhedroses, cytoplasmic polyhedra are being continuously produced in any given infected cell. This continuous process appears to begin apically and to proceed basally. A tendency has been observed for the very smallest polyhedra, at the limit of microscopical resolution in size, to form in batches close to each other and separating as they grow.

(b) Location in Tissues. As previously mentioned, the cytoplasmic polyhedra are associated with quite different disease symptoms from those arising from a nuclear infection. They occur in the cells of the midgut, instead of in the epidermis, tracheae and fat body, as is usual with the nuclear polyhedra. The disease may affect all the cells of the midgut, including the small basal replacement cells, but does not extend to the fore and hind guts.

(c) Reactions with alkali, staining properties, etc. We have seen that the majority of nuclear polyhedra dissolve readily in weak sodium carbonate leaving behind, in many cases, a membrane containing single virus rods or bundles of rods. The cytoplasmic polyhedra react in quite a different manner. They do not dissolve completely but only partially, leaving behind a shell filled with holes rather like a honeycomb (Fig. 3). They appear, on the whole, to be more resistant to the action of alkali than the nuclear polyhedra. There is no membrane and no virus rods have been observed on the electron microscope. On the other hand, in the polyhedra of *Arctia villica* and *A. caja* spherical virus bodies have been found. On many occassions, however, with other cytoplasmic polyhedra it has not been possible to observe the virus by means of the sodium carbonate technique. The reason for this is thought to be that the cytoplasmic polyhedra, or at least parts of them, are more resistant to the action of alkali than the virus contained within them. This latter apparently easily disintegrates leaving the numerous empty holes which give the polyhedra their honeycomb-like appearance. Further evidence on this point will be found in the section dealing with the formation of the virus in the cytoplasm (see also Fig. 3). Some cytoplasmic polyhedra appear break down into thin plates shaped like a horseshoe and it is possible in these cases that the holes are formed by the overlapping of these plates.

The standard method of diagnosing a polyhedral disease is to make a smear of the blood and tissues of the caterpillar on a slide, fix with heat and stain with methylene blue or Giemsa solution. Under these conditions the nuclear polyhedra do not stain but stand out clearly defined against the stained background.

During a study of a polyhedral disease of the larva of *Sphinx ligustri*, the privet hawk moth, (Smith, Wyckoff, and Xeros 1953) it was observed that some polyhedra reacted in a different manner from others, in that they picked up the stain readily and were thus sharply differentiated from the non-staining nuclear type. Later investigation showed that these staining-type polyhedra were cytoplasmic in origin and contained spherical virus particles. As already mentioned, they react differently to weak alkali from the nuclear polyhedra and leave behind a honeycomb structure. We have thus a ready method of differentiating between the two kinds of polyhedra (Fig. 1).

In the bromophenol blue-staining technique, the cytoplasmic polyhedra again react differently from the nuclear type, and the reactions are not always the same with different cytoplasmic diseases. In the case of *T. pellionella*, the case-bearing clothes moth, the cytoplasmic polyhedra stained intensely without pre-treatment with hydrochloric acid. On the other hand, cytoplasmic polyhedra from *P. meticulosa*, the angleshades moth, did not stain to any appreciable extent without pre-treatment with hydrochloric acid (Smith and Xeros 1953 b). So far it has not been possible to see the virus particles inside the cytoplasmic polyhedra on the optical microscope in the same way as they can be observed inside the nuclear polyhedra. Since it is difficult to observe the virus bodies associated with the cytoplasmic polyhedra by means of the sodium carbonate technique another method was used. Thin sections of fixed larvae infected with a cytoplasmic disease were cut and examined on the electron microscope. Various larvae were examined in this way but the most successful sections were obtained with the larvae of *Arctia caja*, the garden tiger moth. Only the gut was sectioned and this revealed an interesting state of affairs. Under high magnification each of these spherical bodies could be seen to be composite, consisting of a number, usually four, of very small units each about 15 mμ in diameter (Figs. 10, 11). The composite bodies were also to be found loose in the cytoplasm, together with a certain number of the small single units. During the course of the disease, the small spheres apparently arise first in the cytoplasm and secondarily associate into the larger bodies prior to the formation of the polyhedra. The minimal infectious unit may well be the small particles of 12–15 mμ diameter. The larger bodies appear to dissociate readily into their small component spheres which, on account of their small size, are easily overlooked (SMITH and XEROS 1954 b, and SMITH 1956).

XEROS (1956) has studied the development of virogenic stromata in the cytoplasmic polyhedroses. These appear to arise subapically in the infected cells. Each virogenic stroma is a definite structural entity with a micro-net structure. In the cords of the stromata spherical virus bodies arise; after their formation the cord material around them disrupts and liberates them into the larger pores formed as a result of the dissolution of the cords. The virogenic stromata grow considerably (up to 10 $\mu \times 2\ \mu$) before small polyhedra form, distributed irregularly over their surfaces. The freed virus bodies become occluded in the polyhedra.

Double Virus Infections

The discovery in diseased larvae of *S. ligustri*, the privet hawk moth, of the staining or cytoplasmic type of polyhedra, together with the non-staining nuclear polyhedra, was the first demonstration of a virus *complex* in an insect (Fig. 1). Since then from studies of a larger number of polyhedroses it has become plain that it is rarely that it can be said definitely that only one type of polyhedral body is present. Even in those cases where a caterpillar develops a predominantly nuclear or cytoplasmic disease, it is usually possible to identify in a smear polyhedra of the other type. It is not surprising, therefore, that inoculation from a caterpillar with a cytoplasmic disease evokes a nuclear disease in the experimental larvae and vice-versa. On the other hand, in the case of *Operophtera brumata*, the winter moth, it is possible to stimulate development of a cytoplasmic polyhedrosis by inoculating the larvae with a nuclear polyhedral virus only (SMITH and RIVERS 1956).

Virus Infection of Pupae and Adults

Although the polyhedral viruses mainly attack the larval forms, it is not uncommon for late-infected larvae to pupate and the disease then develops in the pupa. This has been frequently observed in the case of late-infected larvae of the clothes moth, *Tinea bisselliella*, and pupal infection was observed as long ago as 1914 by WILDERMUTH in a polyhedral disease of the alfalfa caterpillar, *Colias philodice eurytheme* Bdvl. Infected pupae seldom seem to complete their development.

An interesting case of an adult insect developing a polyhedral disease has been observed at Cambridge. Of a stock of larvae of the privet hawk moth, *Sphinx ligustri,* a number died from a natural infection with a cytoplasmic polyhedral disease. The rest survived, and pupated normally, and the adults emerged the following spring. One of these ,a female, died a few days after emergence. In the abdomen were found large quantities of cytoplasmic polyhedra with, apparently, a few nuclear polyhedra, as well.

Polyhedral Disease of a Dipterous Larva

There is a reference in the literature to a polyhedral disease of the larva of the bluebottle fly, *Calliphora vomitoria* (CHAPMAN and

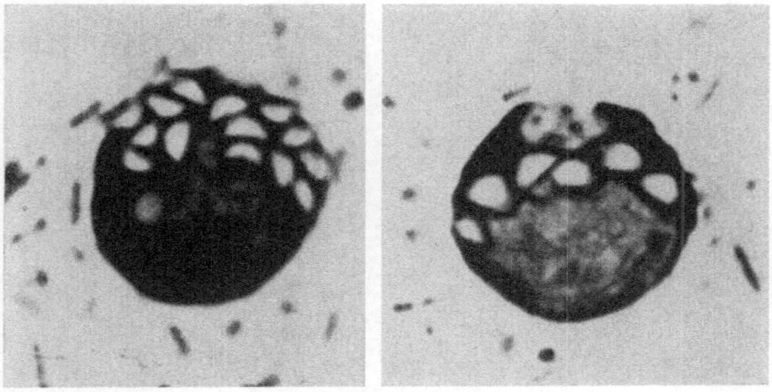

Fig. 5. Two blood cells from a larva of *Tipula paludosa*, the crane fly, infected with a nuclear polyhedral disease; note the "polyhedral" crystals. ×1,750.

GLASER 1915), but this observation has never been repeated. In 1923, RENNIE described briefly a polyhedral disease of the crane fly larva or "leatherjacket," (*Tipula paludosa*). This has now been re-discovered by SMITH and XEROS (1954 c) and has been investigated by modern methods.

Infected larvae are readily identified by their characteristic pallor and when the integument is punctured a milky fluid containing large numbers of "polyhedra" is liberated. Actually the shape of the crystals is not polyhedral but more usually they are shaped like a half moon or the segment of an orange (Fig. 5). The disease is a nuclear one and at the onset a chromatic mass forms in the enlarged nuclei of the blood cells, and from this mass the chromatic material segregates as several spherical bodies, leaving a large eosinophile body in the nucleus. The polyhedra seem to arise around the periphery and are closely applied to the nuclear membrane. They are negatively birefringent and appear to be genuine crystals. Their behaviour in the presence of different reagents is most unusual and differentiates them sharply from either of the other groups of polyhedra. They are resistant to tryspin, and to dilute and weak acids and alkalis. In 1 N sodium hydroxide they elongate to six or more times their length, becoming first biconvex spindles and then elongating into crescent or worm-like shapes (Fig. 6). At about three times their normal length this elongation is still completely reversible, and in water, at pH 5.8, they return to their original size and shape. After such treatment, however, the polyhedra are "activated"; in other words, they now respond in a similar manner in ammonia, 1–2 per cent. sodium carbonate and hydrochloric acid, pH 1–4, but not to 1 N hydrochloric acid or 25 per cent.

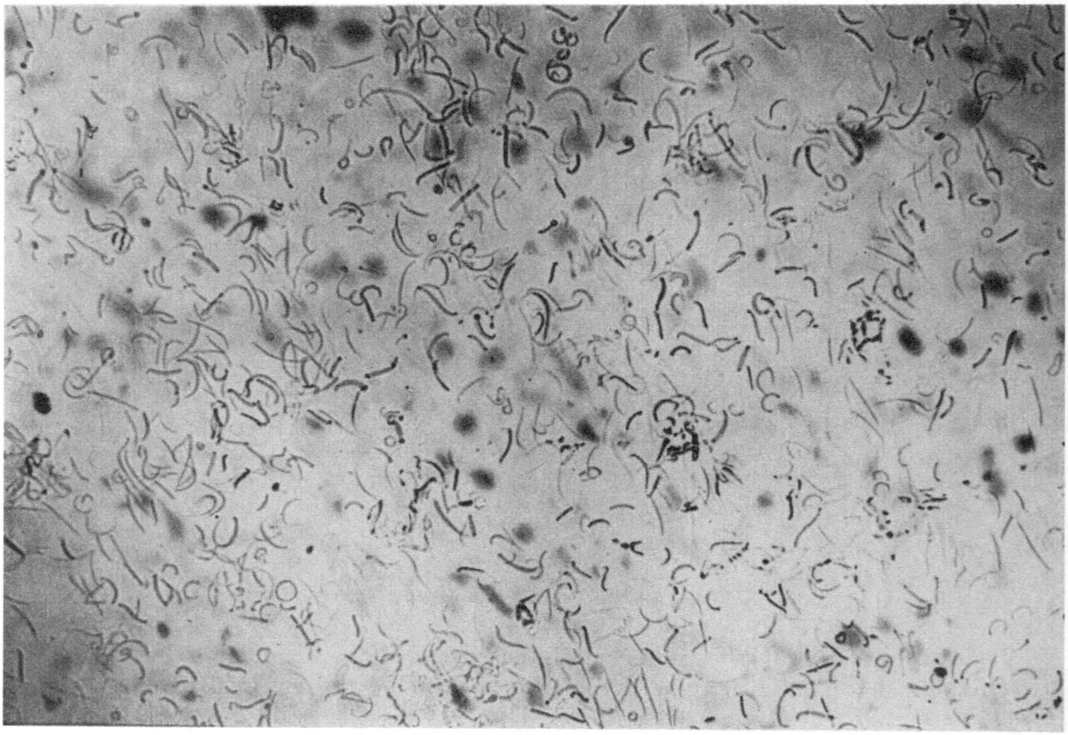

Fig. 6. Polyhedra from *Tipula paludosa* after treatment with ½ N sodium hydroxide and ½ N potassium cyanide for 90 sec. Note the peculiar elongation of the crystals. ×150. (Smith and Xeros.)

sodium carbonate. The elongation and contraction or return to normal shape, which occur along the same axis, can be repeated indefinitely in these solutions, and take place as rapidly as the solutions can be alternated.

The polyhedra are not usually dissolved even after half an hour in 1 N sodium hydroxide at 20° C.

In a solution of equal parts of 1 N sodium hydroxide and 1 N potassium cyanide, the response is speeded up enormously and the polyhedra elongate even further to reach their maximum in 1½ min. They dissolve completely in 2–4 min. The polyhedra give a positive reaction with xanthoproteic and nin-hydrin tests, and stain with bromophenol blue after treatment for 15 min. with 1 N hydrochloric acid at 60° C. Their substance is

Feulgen-positive throughout and each polyhedron has intensely staining granular bands at about ½ μ intervals. Tests for sulphur have been negative.

In such resistant polyhedra as these, it is a difficult matter to observe any possible virus bodies within by means of treatment with alkali. However, studies have been made, on the electron microscope, of polyhedra

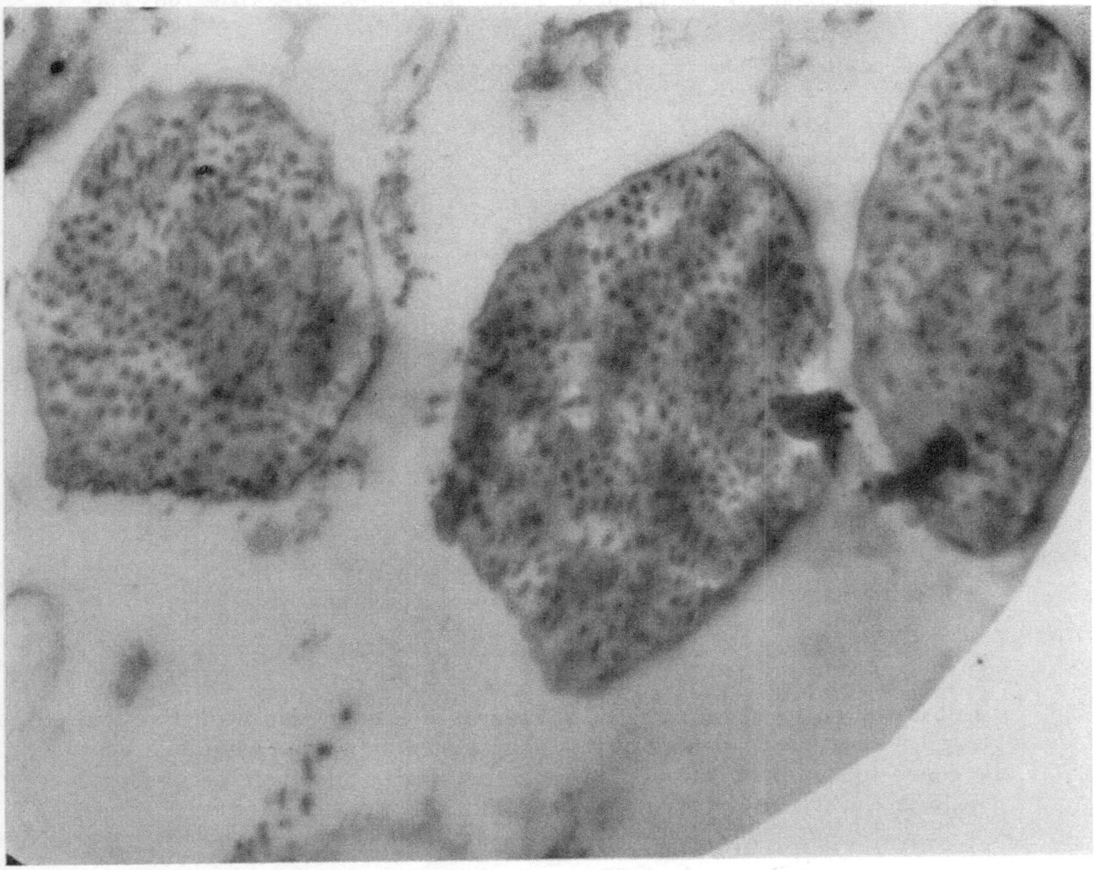

Fig. 7. Ultra-thin section through three polyhedra from the larva of *Tipula paludosa*; note the rod-shaped virus particles inside, some cut transversely and some longitudinally. ×30,000.

subjected to various treatments. In one, the polyhedra were exposed to 1 N sodium hydroxide and washed. The resulting elongated forms appeared semi-opaque with a granular and fibrous structure. In another series, the polyhedra were treated with 1 N sodium hydroxide for 4 min., then washed and treated with sodium thioglycolate at pH 8.4 for 4 min., and finally washed and treated for one minute with 1 N sodium hydroxide. Under this treatment, most of the substance of the polyhedral crystal is dissolved, leaving behind a lens-shaped "ghost." In many polyhedra treated in this fashion rod-shaped bodies have been observed which resembled virus rods. The problem of whether in fact virus bodies were inside the polyhedra was finally solved by cutting thin sections of the crystals. These showed large numbers of short rather fat virus rods embedded in the polyhedral protein (Figs. 7 and 8).

By means of ultra-thin sections for electron microscopy some peculiar changes have been observed taking place inside the nucleus of the virus-infected blood cells of the larvae of *Tipula paludosa*. The virus rods may

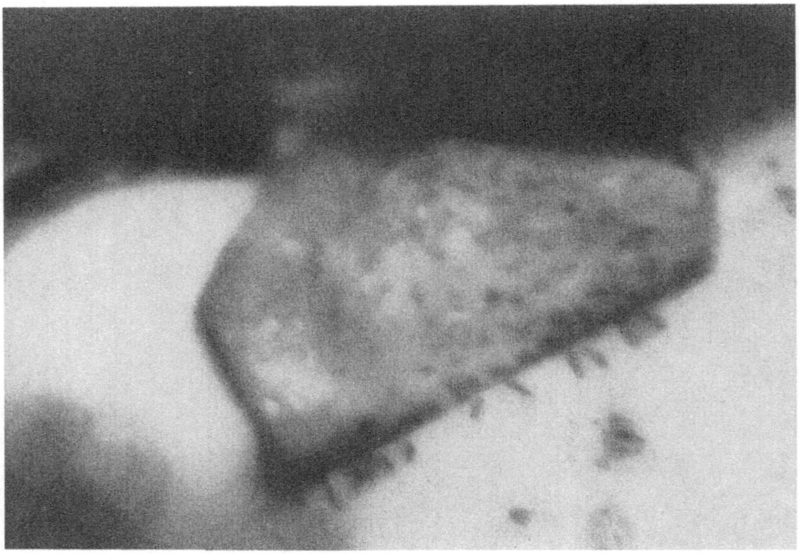

Fig. 8. Similar to Fig. 7; note the virus rods attached end-on to the surface of the crystal. ×30,000.

be observed enclosed each one in the centre of a large thin-walled vesicle (Fig. 9). The exact meaning of this phenomenon is not known, but it may be connected with the formation of the polyhedra and their peculiar extensive nature (SMITH, 1955 a).

Polyhedral Diseases of Hymenoptera

Several polyhedral diseases of the Hymenoptera have been reported but they appear so far to be confined to the sawflies. BALCH and BIRD (1944) have studied a polyhedrosis of the European spruce sawfly, *Gilpinia hercyniae* (Htg.). Unlike the nuclear polyhedroses of Lepidoptera, but similar to the cytoplasmic polyhedroses, the pathological process is concerned with the digestive tract and usually results in the complete destruction of the midgut. The cells of the epithelium of the mesenteron become enlarged and the polyhedral bodies are formed within the enlarged nuclei. The average polyhedron measures 1.3 μ in diameter, ranging from 0.5 to 1.8 μ.

According to Bird (1953), the prepupal stage of the European spruce sawfly is immune to infection by polyhedral disease. The virus multiplies

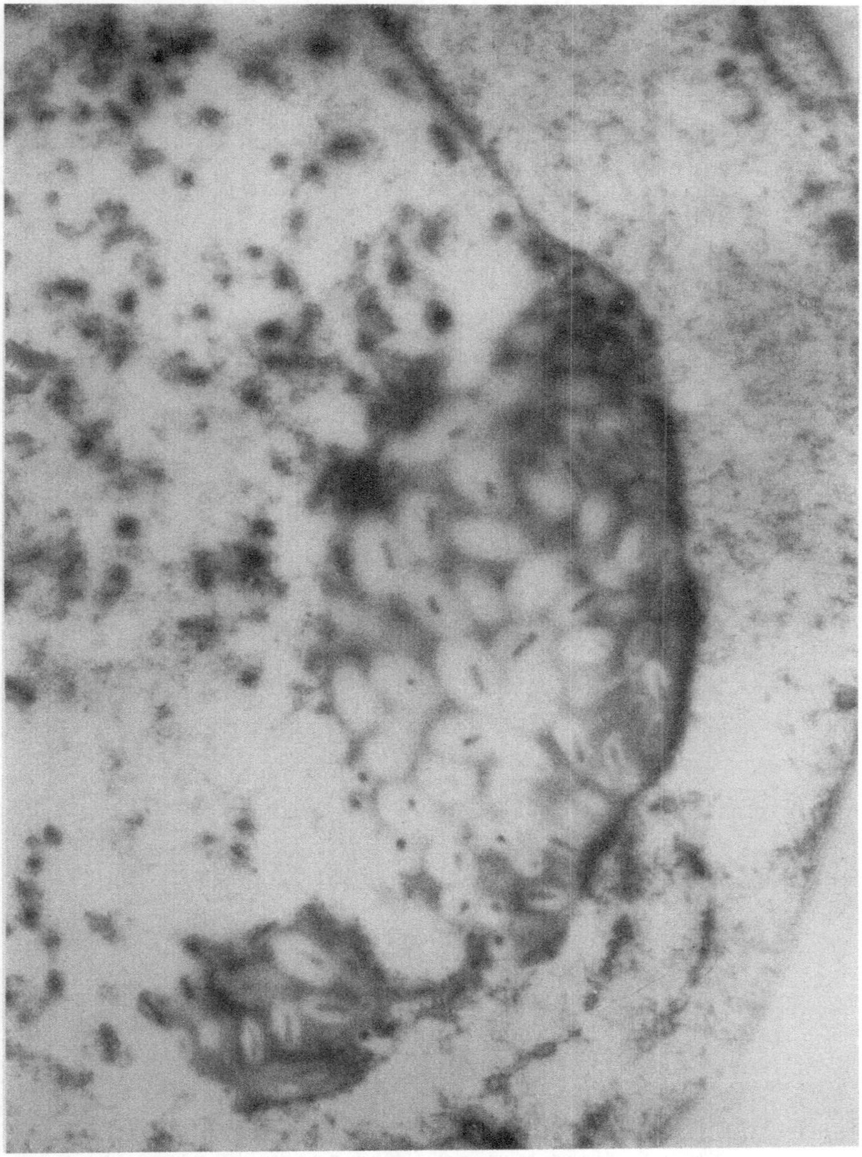

Fig. 9. Ultra-thin section of part of an infected nucleus from the larva of *Tipula paludosa*; note the virus particles enclosed each in a thin-walled vesicle. × 42,000.

only in the nuclei of the digestive cells of the mid-gut epithelium and during metamorphosis these are replaced by embryonic cells not susceptible to infection although they develop into digestive cells eventually. The prepupal mid-gut, a temporary gut, is composed entirely of embryonic cells. New digestive cells appear later in the pupal and adult stages, and these rapidly become infected with virus.

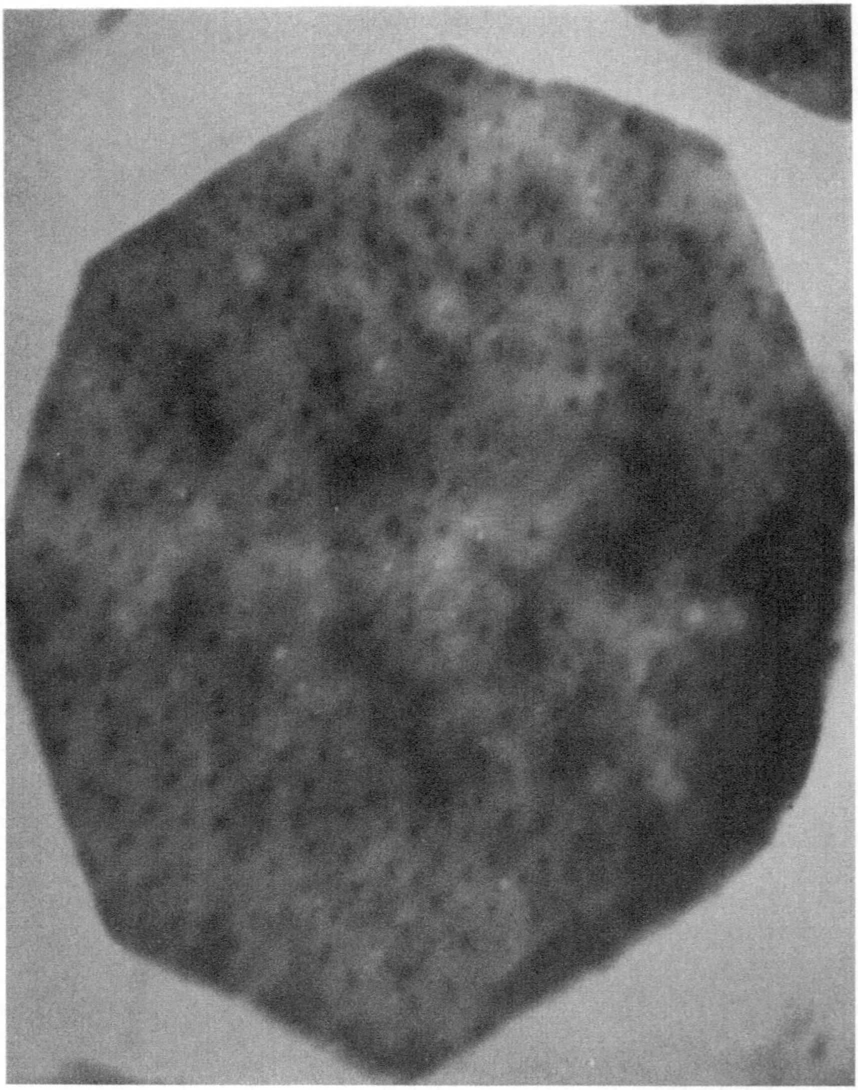

Fig. 10. Ultra-thin section through a cytoplasmic polyhedral crystal from a larva of *Arctia caja*, the garden tiger moth, showing the numerous apparent virus particles. × 60,000.

Bird and Whalen (1953) have described a polyhedral disease of another sawfly larva, the European pine sawfly, *Neodiprion sertifer* (Geoffr.). In this species, as in the European spruce sawfly, polyhedra were found only in the nuclei of the digestive cells of the mid-gut epithelium. The first symptoms of infection are a swelling of the nucleus of the digestive cell and coagulation of the chromatic material of the nucleus. Polyhedra, which appear as small granules, increase in size and cause a further swelling of

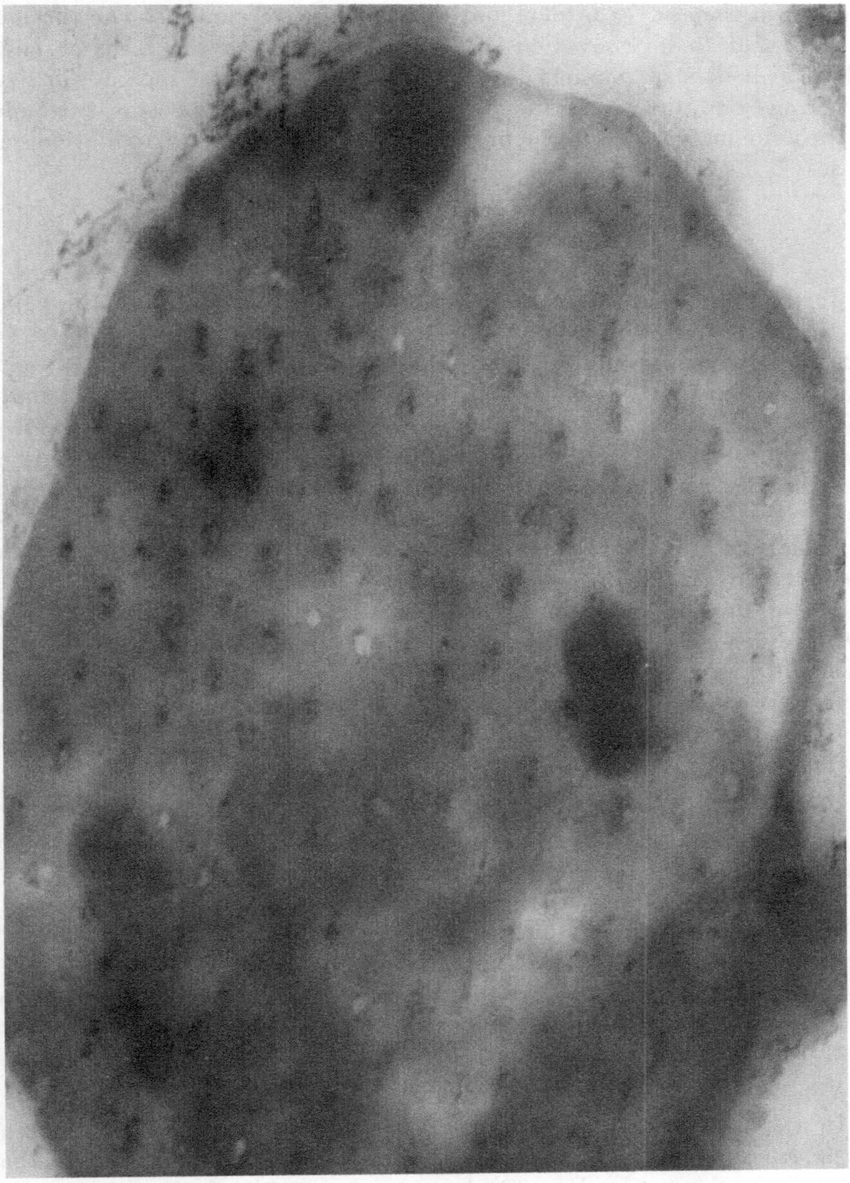

Fig. 11. Similar to Fig. 10 but a thinner section; note the composite nature of the virus particles. ×120,000. (SMITH, J. Biophys. Biochem. Cyt. 1956.)

the nucleus. The polyhedra average about 1 μ in diameter and are frequently almost spherical. They contain mostly rod-shaped virus particles about 250×5 mμ and also some spherical particles.

In a recent paper on the polyhedrosis of a sawfly, *Neodiprion pratti banksianae* Roh., BIRD (1957) suggests that the virus undergoes a life-cycle. Rod-shaped virus particles escape from their developmental membranes and attach themselves by one end to strands of chromatin. The chromatin is then said to be converted into virus particles in the form of minute spherical bodies surrounded by membranes. These later increase in size to form rods. Mature rods and developing stages may become surrounded by protein material to form polyhedra, in which case virus development ceases.

II. Granuloses or Capsular Diseases

In the granuloses the characteristic inclusion consists of an outer capsule or granule within which is another capsule containing a rod-shaped virus particle. In some granuloses there may be more than one virus particle in the inner capsule. The virus particles appear to be always rod-shaped. The first disease of this group to be discovered seems to have been one attacking the larvae of the large white butterfly, *Pieris brassicae* and was described by PAILLOT in 1924. The same worker (1934) described capsular diseases in the larvae of the noctuid moth, *Euxoa segetum.* STEINHAUS (1947) reported a similar infection in the variegated cutworm *Peridroma margaritosa* (Haw.) and in 1948 BERGOLD observed the same type of disease in larvae of the fir-shoot roller, *Cacoecia murinana* (Hb). SMITH and RIVERS (1956) describe six granulosis viruses affecting lepidopterous larvae of economic importance. Since that date the writer has isolated two more granulosis viruses from some African noctuid larvae. It seems probable therefore that this type of virus will eventually prove to be as widespread as the polyhedral viruses.

Larvae of *Pieris brassicae* in various stages of the granulosis disease have been examined by means of sections under the optical microscope. They were fixed in Carnoy's solution and stained with iron haematoxylin. Studies of sections of caterpillars at these different stages of infection seem to support PAILLOT's suggestion that the granules occur in the hypodermis and fat-body. Furthermore, it seems as if the virus develops in the nucleus rather than in the cytoplasm. In sections through the fat-body of a young larva of *P. brassicae,* 96 hours after infection the nuclei appear to be full of granules and there are indications that they rupture, liberating a mass of granules into the cytoplasm of the cell.

In a similar disease affecting the larvae of the noctuid moth *Melanchra persicariae,* the dot moth, the site of multiplication of the virus is almost certainly nuclear. In Fig. 12 the granules inside the greatly enlarged nucleus can be clearly seen and from one nucleus the granules appear to be spilling out in large quantities into the cytoplasm. The hypodermis is similarly affected and the cells become filled with granules (SMITH and RIVERS 1956).

When the granules are treated with weak sodium carbonate and viewed on the electron microscope, the outer granule is seen to be partially dissolved and to have collapsed revealing an inner rod-shaped capsule containing the virus rod (Fig. 13). Occasionally, this inner capsule may also be dissolved and the virus particle itself comes into view. The treatment with sodium carbonate can be carried even further than this so that the

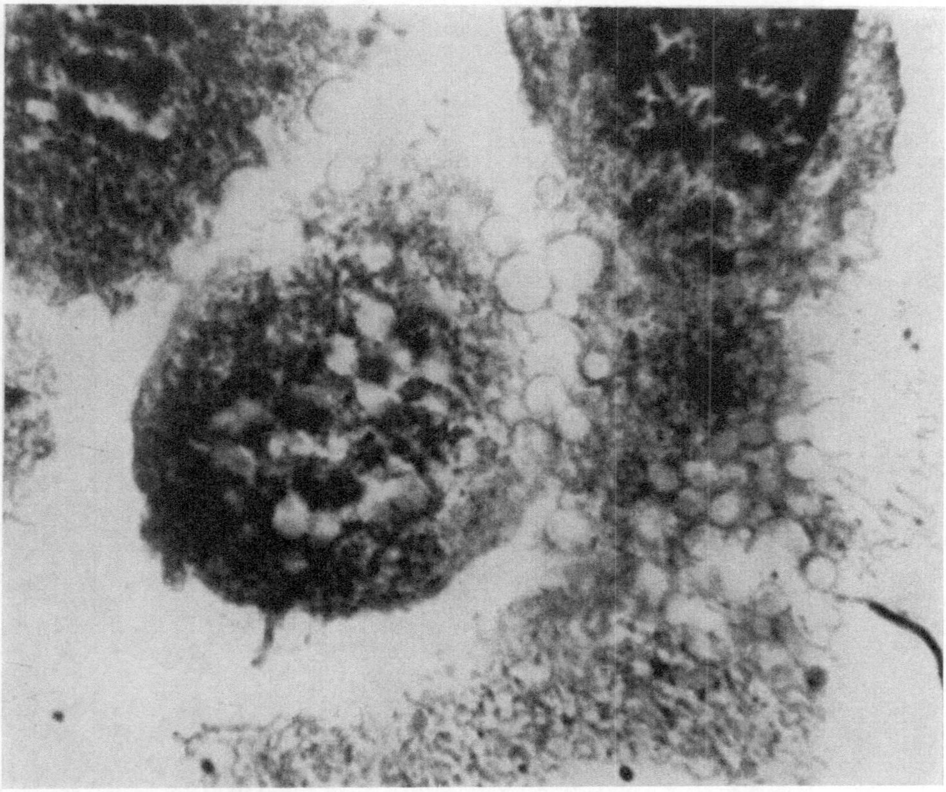

Fig. 12. Section through the fat-body of a larva of *Melanchra persicariae*, infected with a granulosis virus; note the nucleus on the right, apparently liberating granules into the cytoplasm. ×1,500.

contents of the particle itself are partly dissolved away leaving the intimate membrane behind.

So far the granulosis viruses have been reported only from lepidopterous larvae.

III. Viruses without Polyhedral or Granular Inclusions

There are only one or two viruses so far known in this group, one described by WASSER (1952) from *Cirphis unipuncta* (Haworth) the cosmopolitan army worm and the other a new virus from the larva of a fly *Tipula paludosa*, the crane fly (XEROS 1954, SMITH 1954) (Fig. 14). It is with the last-named that we are concerned in this article, and it has some characters which make it unique in the virus world. Multiplication of the virus appears to take place in the cytoplasm of the fat-body cells and infected larvae are readily recognised by the bright blue or violet colour of

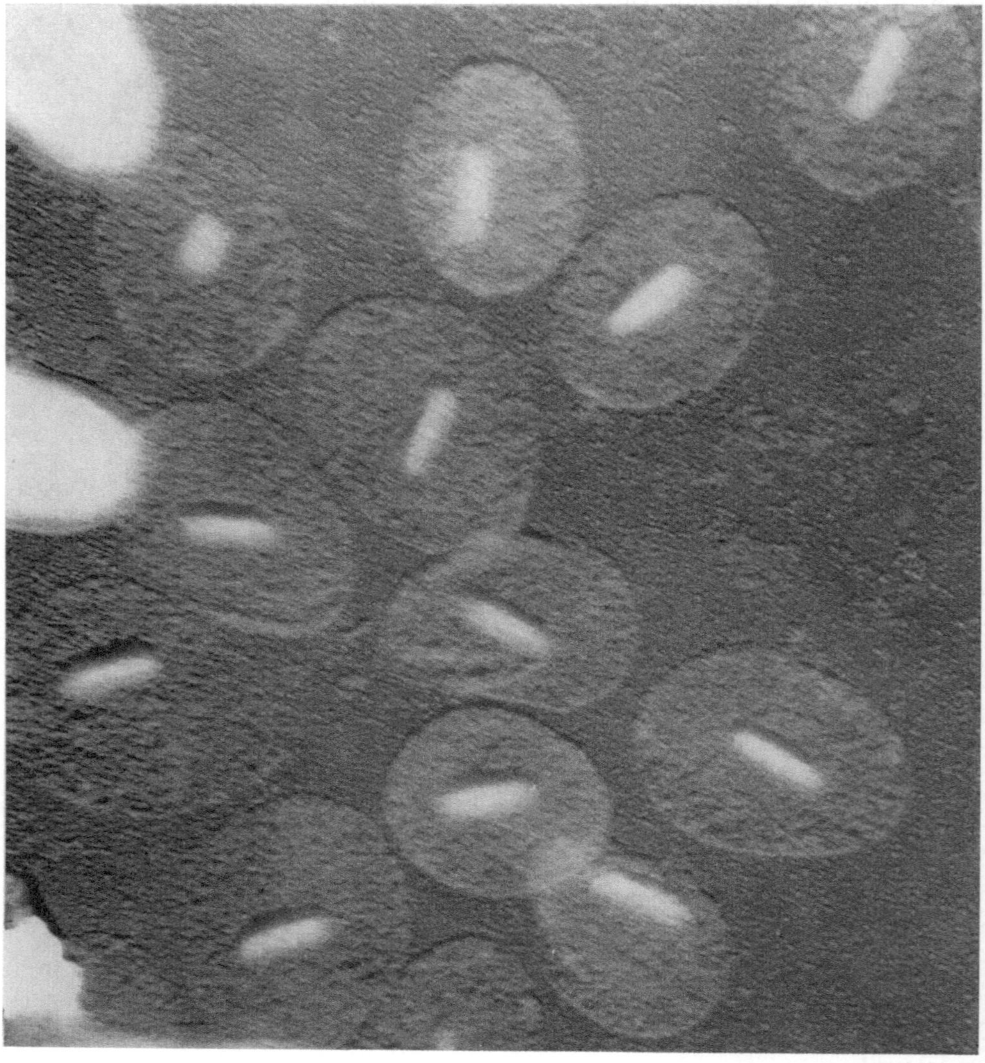

Fig. 13. Granules from a larva of a noctuid moth, *Prodenia lituria*, after treatment with weak sodium carbonate; note collapsed outer granule and the rod-shaped capsule within, which contains the virus particle. ×48,400.

the fat-body showing through the skin. The reasons for this colour are dealt with later. Sections of infected fat-body when observed on the electron microscope show enormous numbers of virus particles so many that almost the whole of the organ appears to be converted into virus (SMITH 1955 b). Indeed, one quarter of the total body weight of the infected larva is virus. The arrangement of the virus particles inside the fat-body is very interesting; in addition to the large numbers of isolated particles, there occur numerous bodies which bear a superficial resemblance

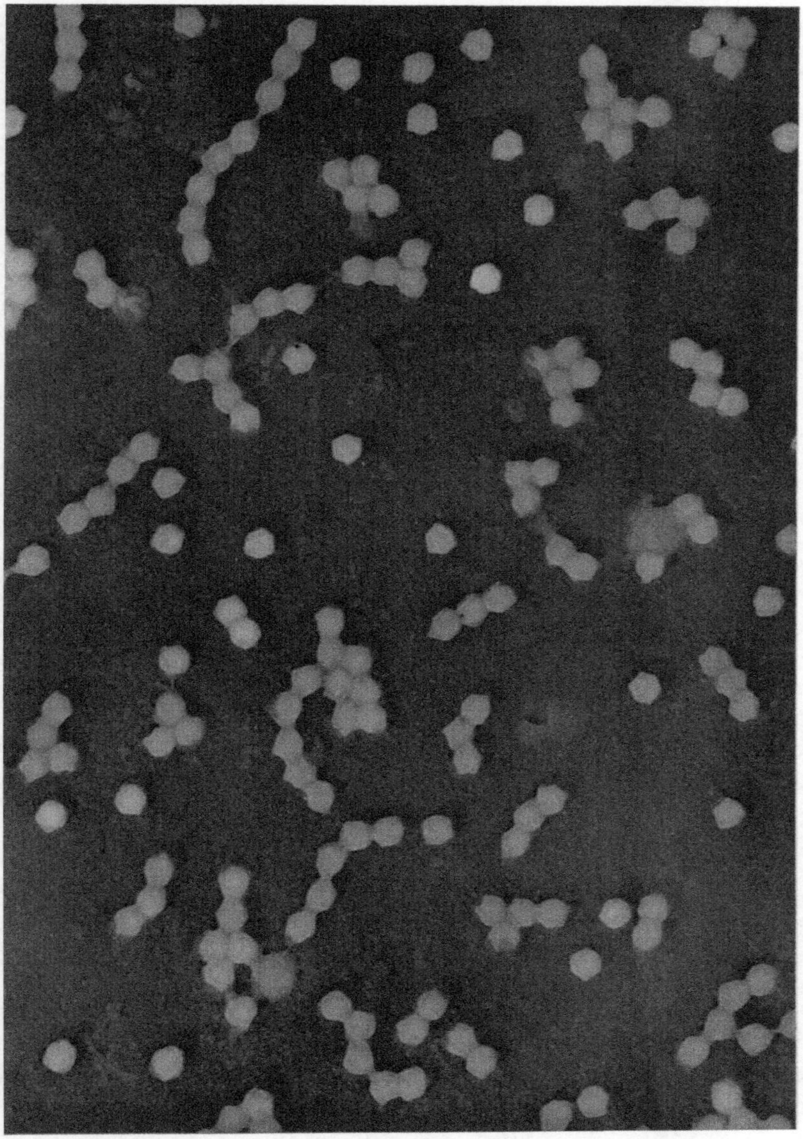

Fig. 14. Particles of a new virus which attacks the larva of the crane fly; note the 5- and 6-sided shape of the particles. ×36,000.

to the intracellular inclusions of other insect viruses. These bodies, however, consist only of virus particles although they do appear to be surrounded by a membrane. A small inclusion body of this type containing a few virus particles is shown in Fig. 15. The virus is rather large but despite this, it is quite uniform in size with a distinctive shape which appears 5- or 6-sided. Since it is large and uniform in size, the virus may be assembled into arrays that exhibit unusual optical properties. When the

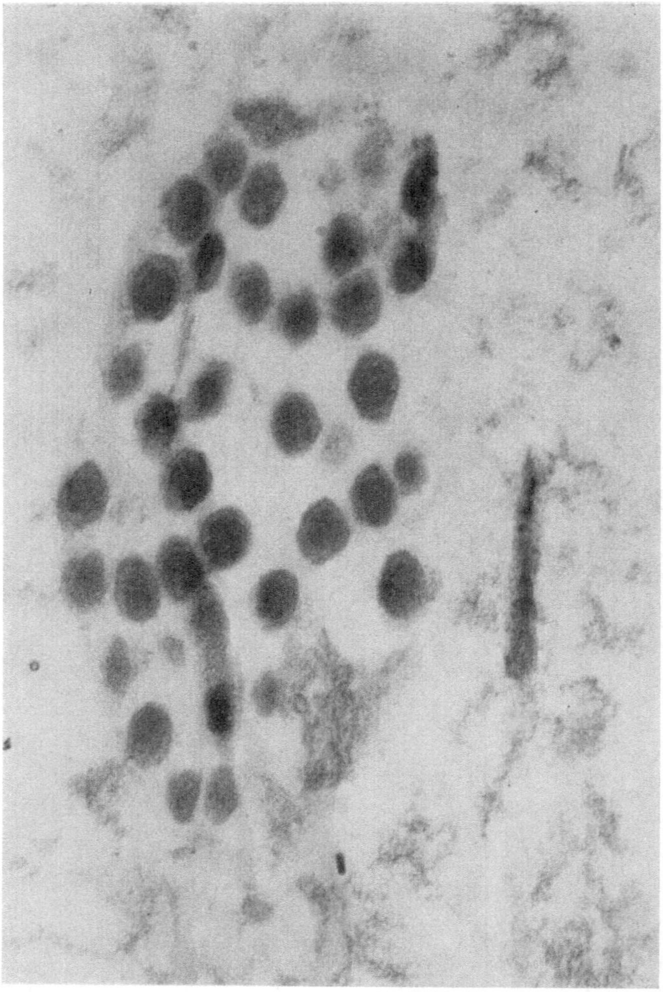

Fig. 15. Ultra-thin section through a small inclusion of the same virus shown in Fig. 14; note the six-sided virus particles with some differentiation of the internal structure. ×64,800.

virus particles are concentrated into a pellet by centrifugation they form into a mass of small crystals showing fascinating colours. The pellet is iridescent and appears somewhat like a large opal. Suspensions of the virus spontaneously crystallize and each crystal is brilliantly coloured when seen by reflected light (WILLIAMS and SMITH 1957). The blue colour of the infected fat-body tissue to which reference has already been made may be accounted for by the high concentration of the virus within, causing the virus particles to be arrayed in a crystalline state. This is the only

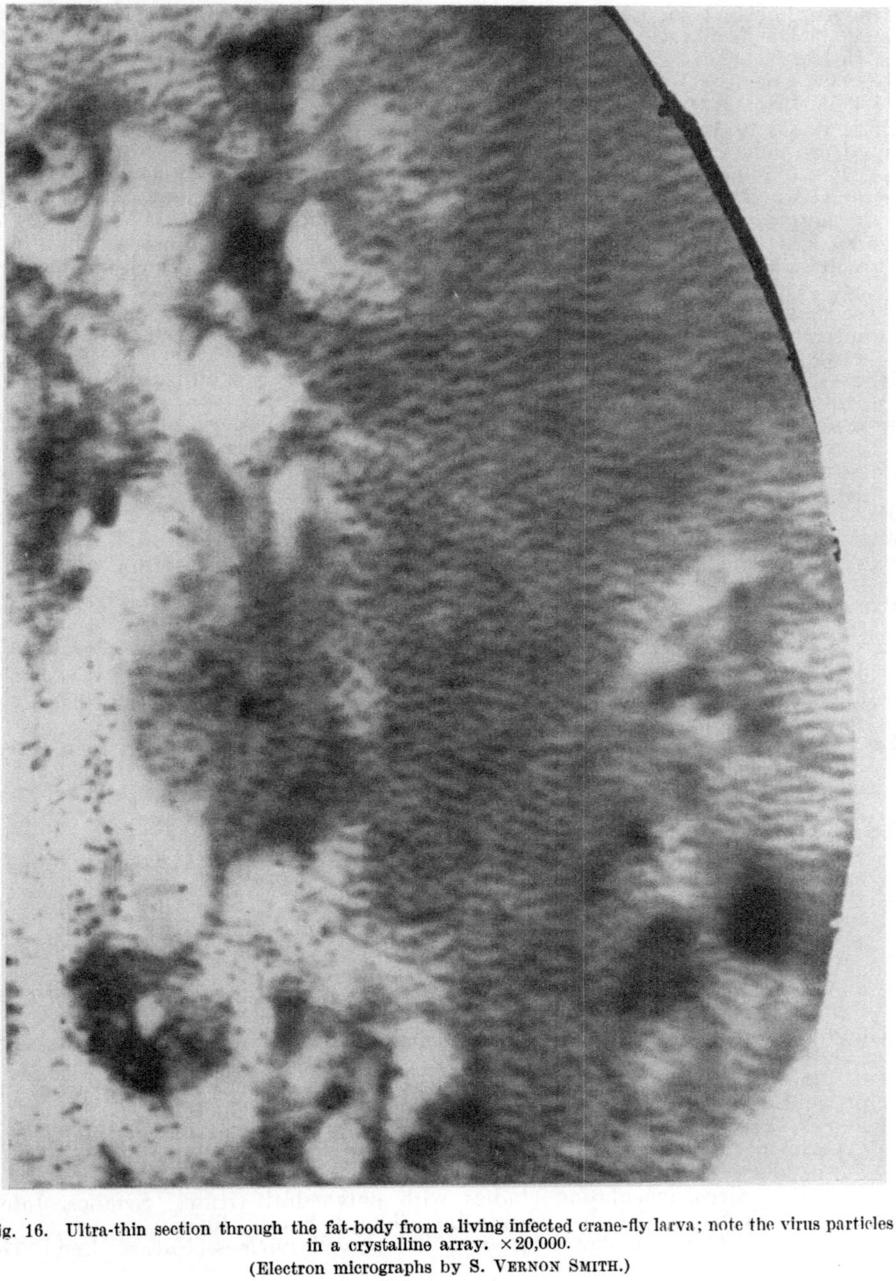

Fig. 16. Ultra-thin section through the fat-body from a living infected crane-fly larva; note the virus particles in a crystalline array. ×20,000.
(Electron micrographs by S. VERNON SMITH.)

known case of a living animal containing a virus in crystalline form within its body (Fig. 16).

References

BALCH, R. E., and F. T. BIRD, 1944: A disease of the European Spruce Sawfly, *Gilpina hercyniae* and its place in natural control. Sci. Agric. **25**, 65—80.

BERGOLD, G. H., 1943: Über Polyederkrankheiten bei Insekten. Biol. Zbl. **63**, 1.

— 1947: Die Isolierung des Polyeder-Virus und die Natur der Polyeder. Z. Naturforsch. **26**, 122.

— 1948: Über die Kapselvirus-Krankheit. Z. Naturforsch. **3 b**, 338—342.

— 1953: Insect Viruses, in Advances in Virus Research, Academic Press, New York.

BIRD, F. T., 1953: The effect of metamorphosis on the multiplication of an insect virus. Canad. J. Zool. **31**, 300—303.

— 1957: On the development of insect viruses. Virology **3**, 237—242.

— and M. M. WHALEN, 1953: A virus disease of the European pine sawfly, *Neodiprion sertifer* (Geoffr.). Canad. Entomol. **85**, 433—437.

CHAPMAN, J. W., and R. W. GLASER, 1915: A preliminary list of insects which have wilt, with a comparative study of their polyhedra. J. Econ. Entomol. **8**, 140—150.

DAY, M. F., I. F. B. COMMON, J. L. FARRANT, and C. POTTER, 1953: A polyhedral disease of a pasture caterpillar, *Pterolocera amplicornis* Walker. Austral. J. exper. Biol. a. med. Sci. **6**, 574—579.

HUGHES, K. M., 1953: The development of an insect virus within cells of its host. Hilgardia **22**, 391—406.

KOMAREK, J., and V. BREINDL, 1924: Die Wipfelkrankheit der Nonne und der Erreger derselben. Z. angew. Entomol. **10**, 99.

LOTMAR, R., 1941: Die Polyederkrankheit der Kleidermotte (*Tineola bisselliella*). Mitt. Schweizer. entomol. Ges. **18**, 372.

MAZIS, D., P. A. BREWER, and M. ALFERT, 1953: Biol. Bull. **104**, 57.

MORGAN, C., G. H. BERGOLD, and H. M. ROSE, 1956: Use of serial sections to delineate the structure of *Porthetria dispar* virus in the electron microscope. J. Biophysic a. Biochem. Cytol. **2**, 23—28.

PAILLOT, A., 1924: Sur une nouvelle maladie des chenilles de *Pieris brassicae*. C. r. Acad. Sci. Paris **179**, 1353—1356.

— 1934: Un nouveau type de maladie à ultravirus chez les insectes. C. r. Acad. Sci. Paris **198**, 204—205.

— and A. GRATIA, 1939: Essai d'isolement du virus de la grasserie des vers à soie par l'ultracentrifugation. Arch. ges. Virusforsch. **1**, 120.

PROWAZEK, S. VON, 1907: Clamydozoa II. Gelbsucht der Seidenraupen. Arch. Protistenk, **10**, 358.

RENNIE, J., 1923: Polyhedral disease in *Tipula paludosa* (Meig.). Proc. r. Phys. Soc. Edin. **20**, 265—267.

SMITH, KENNETH M., 1954: What is a virus? Proc. roy. Instn. **35**, 832.

— 1955 a: Intranuclear changes in the polyhedrosis of *Tipula paludosa* (Meig.). Parasitology **45**, 482—487.

— 1955 b: Morphology and development of insect viruses. In Adv. Vir. Res. III, 199—220, Acad. Press, N.Y.

— 1956: The structure of insect virus particles. J. Biophysic. a. Biochem. Cytol. **2**, 301—306.

— and C. F. RIVERS, 1956: Some viruses affecting insects of economic importance. Parasitology **46**, 235—242.

— and R. W. G. WYCKOFF, 1951: Electron microscopy of insect viruses. Research. Lond. **4**, 148—155.

— — and N. XEROS, 1953: Polyhedral virus diseases affecting the larvae of the privet hawk moth (*Sphinx ligustri*). Parasitology **42**, 287—289.

— and N. XEROS, 1953 a: Studies on the cross-transmission of polyhedral viruses: experiments with a new virus from *Pyrameis cardui*, the painted lady butterfly. Parasitology **43**, 178—185.

— — 1953 b: Cross-inoculation studies with polyhedral viruses. Sympos. Interaction of Viruses and Cells. Istituto Superiore di Sanità, Rome.

— — 1953 c: The development of virus in the cell nucleus. Nature, Lond. **172**, 670.

— — 1954 a: Electron and light microscope studies of the development of the virus rods of insect polyhedroses. Parasitology **44**, 71—80.

— — 1954 b: A comparative study of different types of viruses and their capsules in the polyhedroses and granuloses of insects. Parasitology **44**, 400—406.

SMITH, KENNETH M., and N. XEROS, 1954 c: An unusual virus of a dipterous larva. Nature, Lond. **173**, 866—867.
STEINHAUS, E. A., 1947: A new disease of the variegated cutworm, *Peridroma margaritosa* (Haw.). Science **106**, 323.
WASSER, H. B., 1952: An insect virus without intracellular inclusions. J. Bacter. (Am.) **64**, 787—792.
WELLINGTON, E. F., 1951: Amino acids of two insect viruses. Biochim. et Biophys. Acta **7**, 238.
WILDERMUTH, V. L., 1914: The alfalfa caterpillar (*Eurymus eurytheme*). U. S. D. A. Bur. Entomol. Circ. **133**, 1—14.
WILLIAMS, R. C., and KENNETH M. SMITH, 1957: A crystallizable insect virus. Nature, Lond. **179**, 119—120.
XEROS, N., 1952: Cytoplasmic polyhedral virus diseases. Nature, Lond. **170**, 1073.
— 1954: A second virus disease of the leatherjacket, *Tipula paludosa*. Nature, Lond. **174**, 562.
1956: The virogenic stroma in nuclear and cytoplasmic polyhedroses. Nature, Lond. **178**, 412—413.

Protoplasmatologia
IV. Virus
5. Antibiotika erzeugende virus-ähnliche Faktoren in Bakterien

Antibiotika erzeugende virus-ähnliche Faktoren in Bakterien

Von

Pierre Fredericq

Institut für Mikrobiologie und Hygiene der Universität Lüttich

Inhaltsübersicht

	Seite
I. Colizine und colizinogene Faktoren	1
1. Ausdehnung und Spezifizität der Aktivitätsspektra	2
2. Spezifizität der resistenten Mutanten	3
3. Morphologie der Hemmungshöfe im Agar	3
4. Diffusibilität im Agar	3
5. Thermoresistenz	3
6. Empfindlichkeit gegen proteolytische Enzyme	3
7. Antigene Eigenschaften und elektrophoretische Motilität	3
II. Übertragbarkeit der colizinogenen Faktoren	5
III. Unabhängigkeit der colizinogenen Faktoren von der normalen genetischen Struktur der Bakterien	7
IV. Pathogenität der colizinogenen Faktoren	9
V. Schlußfolgerungen	10
Literatur	12

I. Colizine und colizinogene Faktoren

Manche Coli- oder verwandte Stämme können verschiedene, als Colizin bezeichnete antibiotische Stoffe bilden. Das Colizinbildungsvermögen dieser Bakterien wird durch genetische Faktoren, die in gewisser Hinsicht wie Viren sich verhalten, bestimmt [35, 36].

Der Beweis von Gratia [50] im Jahre 1925, daß der Stamm *E. coli* V einen antibiotischen Stoff ausscheidet, stellt den Ausgangspunkt der Colizinforschungen dar. Dieser antibiotische, Prinzip V genannte Stoff war ein sehr wirksames, sogar in $^1/_{1000}$ Verdünnung, nicht antigenes Agens, das durch Cellophan diffundiert und durch Azeton gefällt wird. Es übte seine Wirkung gegen einen anderen, *Coli φ* genannten *E. coli*-Stamm sowie gegen einen *Sh. dysenteriae*-Stamm aus, hemmte aber nicht das Wachstum anderer daraufhin geprüfter *E. coli*-Stämme [51].

Systematische Untersuchungen haben ergeben, daß der Antagonismus des Typus *Coli* V→*Coli* φ viel häufiger und weniger spezifisch ist, als man zuerst dachte. In der Tat sind ungefähr 50% der *Coli*-Stämme gegen das *Coli* V empfindlich, und ungefähr 25% hemmen das *Coli* φ. Alle die aktiven Stämme wirken jedoch nicht auf dieselbe Weise wie *Coli* V und erzeugen je nachdem einen oder mehrere verschiedene antibiotische Stoffe, die als Colizin A, B, C usw. bezeichnet worden sind. Ganz natürlich bekommt das Prinzip V von GRATIA den Namen Colizin V bei dieser Terminologie [13, 18].

Der Name Colizin bezeichnete ursprünglich die antibiotischen Stoffe, die von *E. coli*-Stämmen ausgeschieden werden. Viele *Shigella* [11, 15], manche *E. freundii* [37] und eigenartige *Salmonella*-Stämme [22, 60] erzeugen jedoch auch antibiotische Stoffe, die mit denen der *E. coli* identisch oder diesen sehr ähnlich sind und so in die Colizingruppe einzureihen sind.

Colizinogene Stämme sind in allen Gegenden der Welt getroffen worden [2, 46, 53, 54, 58, 77, 80]. Die Anwendung von Selektivtechniken hat sie in fast allen menschlichen oder tierischen Stühlen, die ausgeprobt wurden, nachweisen lassen, und zwar von den ersten Lebenstagen an und bei Individuen jeder Altersstufe [52].

Die colizinogenen Stämme weichen von den nichtcolizinogenen Stämmen in keinem anderen Merkmal ab, aber es besteht eine gewisse Beziehung zwischen den biochemischen und antigenen Eigenschaften der Stämme und dem Colizintyp, den sie erzeugen [12, 18]. Die *E. coli*-Stämme können alle Typen bilden, aber die *Shigella*- und *Paracoli*-Stämme erzeugen am häufigsten den Typ E, die *E. freundii*-Stämme den Typ A und die *Salmonella*-Stämme den Typ I.

Es scheint auch, daß gewisse Beziehungen zwischen colizinogenen Stämmen und Darmkrankheiten bestehen. Die ein gegen *Shigella* wirksames Colizin erzeugenden Stämme treten öfter bei Ruhrkranken als bei gesunden Individuen auf [55, 56]. Die das Colizin B, eines der seltenen gegen *S. para* B wirksamen Colizine, erzeugenden Stämme sind bei Paratyphus-B-Kranken sehr häufig [20, 26, 45]. Im Harn von Bauchtyphuskranken findet man beständig einen besonderen *E. freundii*-Typ, der das Colizin A ausscheidet [37].

Die Colizine bilden eine Gruppe von sehr verschiedenartigen antibiotischen Stoffen, die sich voneinander durch zahlreiche Merkmale unterscheiden:

1. Ausdehnung und Spezifizität der Aktivitätsspektra [7, 18, 58]

Die Colizine haben äußerst spezifische Aktivitätsspektra, die streng auf die Stämme der Familie der *Enterobacteriaceae* beschränkt und von einem Colizin zum anderen sehr verschieden sind. Eine Wirkung von colizinogenen Stämmen gegen nicht zu dieser Familie gehörende Keime, die angegeben worden ist [8], muß durch von den Colizinen abweichende Faktoren bestimmt worden sein. Die *E. coli*- und *Sh. sonnei*-Stämme sind häufig gegen zahlreiche Colizine, manchmal gegen alle empfindlich, die *Aerobacter*- und *Salmonella*-Stämme sind seltener und nur gegen gewisse Colizine empfindlich. Die Empfindlichkeit gegen Colizine erlaubt eine Typisierung der verschiedenen Glieder der *Coli-Salmonella*-Gruppe und scheint in Beziehung mit der antigenen Zusammensetzung der Stämme zu stehen [18, 57, 83].

2. Spezifizität der resistenten Mutanten [7, 18]

Die gegen mehrere Colizine empfindlichen Stämme liefern Mutanten, die gegen ein bestimmtes Colizin spezifisch resistent sind, aber gegen die anderen auf den Mutterstamm wirksamen Colizine empfindlich bleiben.

3. Morphologie der Hemmungshöfe im Agar [18]

Wenn man einen colizinogenen Stamm auf eine Agarplatte durch Stich impft und dann 48 Stunden später, nachdem man die erste colizinogene Kultur mit Hilfe von Chloroformdämpfen sterilisiert hat, auf die ganze Oberfläche der Platte einen empfindlichen Indikatorstamm ausbreitet, erhält man Hemmungshöfe, deren Morphologie je nach dem erzeugten Colizintyp verschieden ist. Sie können von jedem Wachstum frei sein, ein ausgedehntes Teilwachstum oder eine wechselnde Zahl von resistenten Kolonien aufweisen. Ihr Rand kann scharf oder progressiv sein, manchmal rosettenförmig, und sich in der Folgezeit mit einer Zone Sekundärlyse umgeben. Die Morphologie der Hemmungshöfe ist vom Typus des erzeugten Colizins und von der Art des gehemmten Stammes abhängig.

4. Diffusibilität im Agar [18]

Die Diffusionsgeschwindigkeit im Agar ist sehr verschieden je nach dem Colizin. Wenn ein Stamm ein sehr diffusibles Colizin ausscheidet, ist der Hemmungshof sehr ausgedehnt, während er mehr beschränkt ist, wenn das Colizin weniger diffusibel ist. Diese Unterschiede stammen wahrscheinlich von Verschiedenheiten in der Molekulargröße her, denn die sehr diffusiblen Colizine können Cellophanmembranen durchdringen, während die anderen zurückgehalten sind.

5. Thermoresistenz [18]

Gewisse Colizine vertragen 100° C Hitze, während andere bei 60—70° geschädigt, sogar vernichtet werden können.

6. Empfindlichkeit gegen proteolytische Enzyme [16, 18]

Die Colizine werden mehr oder weniger schnell durch proteolytische Enzyme wie Trypsin oder verschiedene Bakterienproteasen zerstört.

7. Antigene Eigenschaften [3, 49, 62] und elektrophoretische Motilität [78]

Die mannigfaltigen Eigenschaften der Colizine weisen auf wahrscheinlich sehr verschiedenartige chemische Zusammensetzungen hin. Jedoch zeigen sie alle die allgemeinen Reaktionen der Proteine oder Polypeptide auf [9, 47, 59, 65].

Das Colizin K ist vor kurzem von Goebel und Mitarb. konzentriert und gereinigt worden [48]. Das erhaltene Produkt ist ein farbloser, wasserlöslicher, aus Kohlehydrat, Lipid und Protein zusammengesetzter makromolekularer Stoff, der keine Nukleinsäure enthält, jedoch 6,5% Nitrogen und 1,6% Phosphor. Es zeigt nur eine Komponente in der Elektrophoresis, ist aber heterodispers in der Ultrazentrifugation. Es übt eine mächtige antibiotische Wirkung aus, denn ein Tropfen einer 1 μg/ml Lösung hemmt vollständig das Wachstum empfindlicher Bakterien. Die chemischen, physikalischen, immunologischen und toxischen Eigenschaften dieses Stoffes weisen darauf, daß er das somatische O-Antigen der erzeugenden Bakterien darstellen könnte, aber es ist nicht ausgeschlossen, daß das Colizin K nicht in Wirklichkeit ein getrenntes, an diesen Lipocarbohydratproteinkomplex gebundenes Molekül ist. Die Inaktivierungskurven durch Röntgenstrahlen lassen ein Molekulargewicht zwischen 60.000 und 90.000 erschließen [73].

Die Colizine wirken auf empfindliche Keime durch die Vermittlung von spezifischen Rezeptoren [14, 18]. Ein Stamm, der gegen mehrere verschiedene Colizine empfindlich ist, weist nicht nur einen für seinen Sensibilitätstyp charakteristischen und für alle Colizine, gegen die er empfindlich ist, identischen Angriffspunkt auf, wohl aber mehrere für jedes Colizin spezifische Rezeptoren.

Die Anwesenheit dieser Colizinrezeptoren wird durch genetische Faktoren bestimmt, die durch Kreuzung rekombinieren [38, 71] oder durch Mutation verschwinden [17, 18] können. Diese Mutationen entstehen spontan und unabhängig von den anderen Eigenschaften. Die verschiedenen Colizinrezeptoren sind auch untereinander unabhängig, denn ein Mutant, der gegen ein bestimmtes Colizin resistent ist, bewahrt die für den Stamm, aus dem er hervorgeht, charakteristische Empfindlichkeit gegen die anderen Colizine. Doch kann man durch eine Reihe von aufeinanderfolgenden Mutationen einen ursprünglich gegen zahlreiche Colizine empfindlichen Stamm in einen total resistenten Stamm verwandeln, indem man ihm nacheinander alle Rezeptoren, die er besitzt, wegnimmt.

Die Spezifizität dieser Rezeptoren erlaubt eine Gruppierung der Colizine nach den Rezeptoren, auf denen sie haften, aber diese Gruppen können Colizine mit wahrscheinlich sehr verschiedenen chemischen Zusammensetzungen enthalten, zum Beispiel die Colizine der Gruppe E [19, 26]. Man muß auch beachten, daß diese Rezeptoren nicht immer von einem einzigen Colizin streng spezifisch sind.

Diese Colizinrezeptoren sind faktisch Fixationsrezeptoren [6, 23]. Extrakte von empfindlichen Bakterien fixieren und neutralisieren die Colizine, gegen die diese Bakterien empfindlich sind. Extrakte von resistenten Bakterien haben keine solche Wirkung [5]. Die Anwesenheit dieser Fixationsrezeptoren wird auch durch die Wirkung von Antibakterienseren festgestellt. Diese Seren üben keine eigene Anticolizinwirkung aus, aber sie schützen die empfindlichen Bakterien gegen die spätere Wirkung der Colizine, wahrscheinlich indem sie ihre Rezeptoren maskieren [4].

Die Fixation auf einem spezifischen Rezeptor ist eine unentbehrliche, aber doch unzureichende Bedingung für die Wirkung der Colizine, denn gewisse Stämme können ein Colizin fixieren, obgleich sie unempfindlich sind. Wenn empfindliche Zellen Colizin adsorbieren, werden sie schnell getötet. Die Colizine sind bakterizide, aber nicht bakteriolytische Agenzien. Man kann jedoch in gewissen Fällen eine späte partielle Lyse beobachten, wahrscheinlich infolge einer Autolyse der abgestorbenen Bakterien [18].

Die Kinetik der bakteriziden Wirkung ist bei den Colizinen ML, E und K untersucht worden [23, 43, 69]. Die Proportion der überlebenden Keime steht im direkten Verhältnis zu der Keimzahl, auf die man das Colizin wirken läßt. Sie ist dagegen eine exponentielle Funktion der Wirkungszeit und der Konzentration des Colizins. Die Absterbenskurven drücken also das Adsorptionsphänomen, das der Colizinwirkung zugrunde liegt, aus.

Die bakterizide Wirkung der Colizine führt ausgeprägte Metabolismusabänderungen der empfindlichen Bakterien herbei [69]. Nach Hinzufügen des Colizins ML zu einer Kultur empfindlicher Bakterien hört das Wachstum unverzüglich auf. Die Atmungsintensität bleibt zunächst während ungefähr 20 Minuten unverändert, dann nimmt sie progressiv ab. Die RNS- und DNS-Synthesen sind sofort gestört. Das Colizin scheint also alle Bakteriensynthesen zu blockieren. Es blockiert auch die Vermehrung gewisser Bakteriophagen [27, 69].

Die colizinogenen Eigenschaften sind Erbmerkmale von größter Stabilität. Das im Jahre 1925 von Gratia isolierte *Coli* V erzeugt 30 Jahre später immer noch dasselbe Colizin, und viele während mehr als zehn Jahre untersuchte Stämme haben keine Veränderungen ihrer colizinogenen Eigenschaften aufgewiesen. Über 15 Jahre lang trocken aufbewahrte Stämme liefern Colizin bei der ersten Fortzüchtung [81]. Die Konstanz dieser Eigenschaften macht ihre Benützung als typisierende Merkmale bei epidemiologischen Untersuchungen an *Sh. sonnei*- und Dyspepsiecoli-Stämmen, die häufig colizinogen sind, möglich [1, 42]. Die aus ein und demselben epidemiologischen Herd herkommenden Stämme sind nämlich alle colizinogen bzw. nichtcolizinogen, und wenn sie colizinogen sind, erzeugen sie alle dasselbe Colizin. Jedoch ist die Erbstabilität nicht immer absolut. Die Colizinproduktion kann quantitative Varianten aufweisen und colizinogene Stämme können ausnahmsweise Mutanten liefern, die diese Eigenschaften endgültig verloren haben [18].

Auf Grund ihrer Erbstabilität und ihrer Beständigkeit müssen die colizinogenen Eigenschaften durch genetische Faktoren bedingt sein, die die Colizinsynthese steuern und ihre genetische Kontinuität sicherstellen. Wir werden jetzt feststellen, daß diese genetischen Faktoren übertragbare, von der normalen genetischen Struktur der Bakterien unabhängige und pathogene Agenzien sind.

II. Übertragbarkeit der colizinogenen Faktoren

Die colizinogenen Faktoren können von einem Stamm auf einen anderen übertragen werden [28]. Um diese Übertragung nachzuweisen, ist es notwendig, eine Technik doppelter Selektion anzuwenden. Man impft in ein und dasselbe Bouillonröhrchen einen colizinogenen Stamm, der streptomyzinempfindlich ist, und einen nichtcolizinogenen Stamm, der gegen dieses Antibiotikum resistent ist. Nach 24 Stunden Wachstum breitet man auf der Oberfläche von streptomyzinenthaltenden Agarplatten aufeinanderfolgende Verdünnungen dieser Mischkultur aus, dann gießt man eine zweite Schicht streptomyziniertes Agar darüber. Die Kolonien, die sich zwischen den beiden Agarschichten entwickeln, kommen alle von dem nichtcolizinogenen Stamm her, da der andere durch das Streptomyzin gehemmt wird. Man erfaßt dann unter diesen Kolonien diejenigen, die eventuell colizinogen geworden sind, indem man 48 Stunden später auf der ganzen Fläche der oberen Schicht einen colizinempfindlichen (und natürlich streptomyzinresistenten) Indikatorstamm ausbreitet. Am anderen Tag zeigt die zusammengeflossene Kultur dieses Indikatorstammes kreisförmige Hemmungshöfe, in deren Zentren in der Tiefe die colizinogen gewordenen Kolonien liegen und leicht weiter zu impfen sind.

Die colizinogenen Eigenschaften gewisser *E. coli-* oder selbst *Sh. sonnei*-Stämme können so auf andere nichtcolizinogene Stämme, ebenso auf Stämme anderer Arten, wie *paracoli, K. pneumoniae, S. typhi* bzw. *S. para B,* übertragen werden [28, 63]. Die umgewandelten Stämme bewahren alle die Merkmale, die die ursprünglich nichtcolizinogenen Stämme charakterisieren, und unterscheiden sich von ihnen nur durch die neuerworbene colizinogene Eigenschaft. Zum Beispiel ein durch Mischkultur mit einem colizinogenen *Sh. sonnei*-Stamm colizinogen gemachter *E. coli*-Stamm behält nicht nur die Sensibilitäts- bzw. Resistenzmerkmale, die den ursprünglichen Stamm markieren, sondern auch alle Merkmale, die die *E. coli-* von der *Sh. sonnei*-Art gründlich unterscheiden, namentlich Gasbildung durch Glucosevergärung, Laktose und Xylosespaltung, Indolbildung.

Diese genetische Übertragung eines einzigen Charakterzuges durch Mischkultur zweier Stämme gehört in den Rahmen des von Lederberg [75] als *Transduktion* bezeichneten Phänomens. Dieser Ausdruck wird häufig in dem einschränkenden Sinne einer genetischen Übertragung durch Vermittlung von Phagenpartikeln gebraucht, doch spielen die Phagen keinerlei Rolle in der Transduktion der colizinogenen Eigenschaften [29].

Die Transduktion ist für jedes Colizin spezifisch. Zum Beispiel der durch den colizinogenen CA 18-Stamm transduzierte *E. coli*-K 12-Stamm erzeugt wie CA 18 das Colizin B. Derselbe durch den colizinogenen K 235-Stamm transduzierte *E. coli*-K 12-Stamm erzeugt wie K 235 das Colizin K usw. Bei Anwendung von entsprechenden Selektivtechniken kann eine serienweise Transduktion nachgewiesen werden. Ein schon transduzierter Stamm kann erneut durch einen Stamm, der ein anderes Colizin bildet, transduziert werden und so zwei verschiedene Colizine ausscheiden.

Die Wirksamkeit der Transduktion ist den Fällen nach sehr verschieden. Im allgemeinen stellen die umgewandelten Zellen 1 bis 10% der nichtumgewandelten Zellen dar, aber das Verhältnis kann fast 100% erreichen, wenn der zu transduzierende Stamm gegen das Colizin des transduzierten Stammes sehr empfindlich ist. In diesem Falle sind die nichtcolizinogenen Zellen getötet worden und es entwickeln sich in der Mischkultur neben den anfangs colizinogenen Zellen nur die umgewandelten Zellen, von nun an immun, wie wir es später sehen werden, und gegebenenfalls spontane Resistentmutanten.

Alle Stämme sind nicht unbedingt fähig, ihre colizinogenen Eigenschaften zu transduzieren, vielleicht wegen einer F-sexuellen Polarität, aber andere Faktoren müssen auch eine Rolle spielen, denn Stämme, die mehrere Colizine erzeugen, können im allgemeinen nur eine einzige ihrer colizinogenen Eigenschaften übertragen. Die transduzierende Fähigkeit scheint von dem Colizintyp abzuhängen. Sie ist sehr häufig bei Colizin I erzeugenden, weniger häufig bei Colizin B, E bzw. K erzeugenden und nur ausnahmsweise bei Colizin V erzeugenden Stämmen zu beobachten [33].

Die Übertragung der colizinogenen Eigenschaften tritt äußerst schnell ein, wenigstens anfangs. Es ist nicht notwendig, die zwei Stämme 24 Stunden lang gemeinsam zu züchten, es genügt vielmehr, sie einige Minuten zu vermischen, um schon eine Übertragung zu erzielen. Obwohl die colizinogenen Eigenschaften übertragbar sind, sind die transduzierenden Agenzien in das äußere Milieu nicht entlassen worden. Die Transduktion ist an die Anwesenheit von lebenden colizinogenen Zellen gebunden und konnte nicht durch Wirkung von getöteten colizinogenen Kulturen oder Filtraten, selbst nicht von überstehenden Flüssigkeiten bei der Zentrifugierung lebender Kulturen, erreicht werden.

Die Übertragung scheint die Konjugation zweier Stämme zu fordern, denn die serienweise in *E. coli* K 12 untersuchte Transduktion hat nachgewiesen, daß die transduzierende Wirksamkeit an die sexuelle F_+-Polarität gebunden ist. Ein F_--Stamm kann nicht seine colizinogenen Eigenschaften übertragen, aber es genügt, ihn F_+ zu machen, um die Transduktion zu ermöglichen [30].

III. Unabhängigkeit der colizinogenen Faktoren von der normalen genetischen Struktur der Bakterien

Obwohl die Übertragung der colizinogenen Eigenschaften den Kontakt und wahrscheinlich die Fusion zweier Zellen erfordert, ist sie jedoch von der Rekombination der anderen Erbmerkmale unabhängig.

Das Vorhandensein von Sexualitätserscheinungen bei *E. coli* hat eine Kreuzungsanalyse möglich gemacht. Das Aufzeigen von Zellen, die die Erbcharaktere zweier verschiedener Eltern rekombinieren, setzt die Anwendung von Selektivtechniken, wie Selektion der prototrophen Rekombinanten auf einem synthetischen Nährboden nach Kreuzung komplementärer auxotropher Mutanten, voraus, um sie aus der Masse der aus direkter Teilung der Eltern entstehenden Zellen zu isolieren [74].

Wird eine Mischung zweier diauxotropher Doppelmutanten, zum Beispiel eines biotin- und methioninbedürftigen (B- M-) und eines threonin- und leucinabhängigen (T- L-) Stammes auf einem wirkstofffreien „Minimal"-Boden verbreitet, wachsen nur die von den obigen Nährstoffen unabhängigen prototrophen Kolonien, die eine Neukombination vom Typ $B_+M_+T_+L_+$ darstellen, und können so ausgelesen werden. Diese Neukombination entsteht durch Umkombination der Erbfaktoren der beiden Elterntypen $B_+M_+T_-L_-$ und $B_-M_-T_+L_+$, und zwar durch ein Crossing-over zwischen den beiden Faktoren M und T. Führt man nun weitere Erbunterschiede in die beiden Elternstämme durch Mutationsauslösung (z. B. biochemische Varianten oder Resistenz gegen Phagen und Antibiotika) ein, und zwar solche, die durch den Nährboden nicht wie die Auxotrophiefaktoren selektioniert werden, dann erhält man in den prototrophen Rekombinantenkolonien die verschiedenen Kombinationen dieser weiteren Faktoren in ganz bestimmten Verhältnissen. Die Beständigkeit dieser Verhältnisse ist ein Hinweis darauf, daß die Faktoren in *E. coli* in einer linearen Reihe, kettenartig, zusammengeschlossen sind und daß ein Crossing-over um so häufiger geschieht, je größer die Entfernung zwischen den beiden Faktoren ist.

Die fruchtbaren *E. coli*-Stämme sind jedoch nicht alle in der Lage, mit anderen zu kreuzen. Manche, als F_- bezeichnete Stämme können nur mit anderen, F_+ bezeichneten Stämmen rekombinieren. Diese F_+-Stämme enthalten einen hypothetischen F-Faktor, der durch Transduktion übertragbar ist und verlorengehen kann. Die durch diesen F-Faktor bedingte Sexualpolarität ist besonders wichtig zu beachten, als die Rekombinanten nicht aus der Fusion zweier kompletter Zellen entstehen, sondern durch die Übertragung nur eines Teiles des genetischen Materials von einer F_+-Zelle auf eine vollständige Zelle des F_--Partners. Die Übertragungsfrequenz der verschiedenen Erbfaktoren untersteht so festaufgestellten Gesetzen und wird durch ihre Koppelung mit den selektionierten Markören und durch die F-Polarität der Elternstämme bestimmt [76].

In diesen Kreuzungen kann man nun einen oder den anderen Partner durch Transduktion colizinogen machen. Die colizinogenen Eigenschaften verhalten sich ganz unabhängig von den anderen Eigenschaften und sind an keine anderen Erbmerkmale gebunden. Sind sie von dem F_--Partner getragen, dann werden sie immer, ohne Ausnahme, auf alle Rekombinanten übertragen, welche auch die auswählenden Marköre sein mögen. Die Inversion dieser Eigenschaften, selbst wenn die Sexualpolarität und die selektionierten Marköre bei der Kreuzung dieselbe bleiben, kehrt die Übertragungsfrequenz nicht um, denn zahlreiche Rekombinanten, bis zu 70%, erben noch die colizinogenen Eigenschaften, wenn sie den F_+-Partner charakterisieren. Die aus der Kreuzung zweier verschieden colizinogener Eltern abstammenden Rekombinanten erzeugen alle das Colizin des F_--Partners, jedoch bilden einige, die doppeltcolizinogen sind, auch dasjenige des F_+-Partners [39, 40].

Die colizinogenen Eigenschaften können die Fertilität der Kreuzungen mehr oder weniger deutlich vermindern oder gar aufheben, je nach dem erzeugten Colizintyp. Die Verminderung wird schon deutlich bei den Kreuzungen, bei denen der F_--Partner colizinogen ist, zeigt sich aber vor allem, wenn die colizinogenen Eigenschaften von dem F_+-Partner getragen werden. Wie im Falle der lysogenen Eigenschaften wird die Hemmung wahrscheinlich ausgelöst durch Induktion der Colizinproduktion während der Rekombination und durch Vernichtung der Rekombinanten, die die induzierten colizinogenen Faktoren erhalten. Trotz ihrer Rolle in der scheinbaren Fertilität der Kreuzungen beeinflussen die colizinogenen Eigenschaften die Übertragungsfrequenz anderer Erbfaktoren nicht, was die Abwesenheit eines Linkage bestätigt [41].

IV. Pathogenität der colizinogenen Faktoren

Colizinogene Kulturen sind gegen das besondere Colizin, das sie ausscheiden, nicht empfindlich, obgleich sie unter Umständen durch andere Colizine gehemmt werden können. Diese Immunität ist eine direkte Folge ihrer colizinogenen Eigenschaft, denn nichtcolizinogene Mutanten sind im allgemeinen empfindlich gegen das Colizin, das der Mutterstamm, aus dem sie hervorgehen, erzeugt [18].

Diese Immunität betrifft nur die Kultur, wenn sie als ein Ganzes betrachtet wird, aber nicht die einzelnen Zellen. Colizin wird nicht von allen Zellen einer colizinogenen Kultur ausgeschieden, und die wenigen Zellen, in denen die Colizinsynthese stattfindet, werden durch diese Synthese getötet. Das Verschwinden dieser seltenen Zellen wird durch die Entwicklung der anderen Zellen ausgeglichen, so daß eine colizinogene Kultur scheinbar normal wächst.

Die Colizinsynthese kann jedoch in fast allen Zellen einer colizinogenen Kultur induziert werden, durch zahlreiche mutagene oder cancerigene Agenzien, namentlich UV-Strahlen, Peroxyde, Ethylenimine oder Halogenoalcoylamine [31, 69, 79, 82]. Die Colizininduktion ist zuerst bei dem Stamm *E. coli* ML beobachtet worden und führte zur massiven Lyse dieses Stammes [68]. Jedoch ist dieser ML-Stamm gleichzeitig lysogen und colizinogen, und die Lyse muß eher der Induktion seines Prophagen zugeschrieben werden, denn die Colizininduktion bei anderen Stämmen, die nicht lysogen sind, bringt nicht eine solche Lyse hervor [32, 64, 72]. Die Colizine sind faktisch bakterizide, aber nicht bakteriolytische Agenzien.

Die Induktion der Colizinbildung bei den colizinogenen Stämmen ist der Induktion der Prophagenentwicklung bei den lysogenen Stämmen sehr ähnlich. In beiden Fällen geht dem Tod der Zellen eine ungefähr 90 Minuten lange Residualwachstumsperiode voraus, mit paralleler Steigerung der Atmungsintensität. Jedoch beginnt die Colizinsteigerung früher als die Phagenentlassung, da diese nur nach Lyse der Zellen hervortritt. Colizinogene wie lysogene Stämme verewigen also erblich einen potentiellen Letalfaktor, dessen spontane oder induzierte Aktivierung zum Tod der Zellen führt.

Die ursprünglich nichtcolizinogenen, aber durch Transduktion colizinogen gemachten Stämme verhalten sich ganz wie die spontan colizinogenen Stämme. Das Einführen eines colizinogenen Faktors in die transduzierten Zellen zieht die Immunität gegen das Colizin, dessen Synthese er steuert, nach sich, selbst wenn diese Zellen ursprünglich gegen dieses Colizin empfindlich waren, und seine spontane oder induzierte Entwicklung führt auch zum Tod der Zellen [32, 64].

Die Immunität der colizinogenen Kulturen gegen das Colizin, das sie bilden, ist ganz verschieden und unabhängig von der Resistenz durch Verlust der Colizinrezeptoren [34]:

1. Die durch Transduktion colizinogen gemachten Kulturen behalten den Rezeptor für das Colizin, das sie jetzt ausscheiden, wenn sie ihn vor der Transduktion enthielten. Die Gruppe E zum Beispiel umfaßt Colizine, die wahrscheinlich chemisch sehr verschieden sind, die aber alle wirken, indem sie auf demselben Rezeptor der empfindlichen Bakterien haften, auf dem sich auch der Bakteriophage BF 23 fixiert. Die für ein Colizin der Gruppe E colizinogen gemachten Stämme bewahren diesen gemeinsamen Rezeptor und sind nur gegen das besondere Colizin, das sie bilden, immun. Dagegen sind die Mutanten, die diesen Rezeptor verloren haben, immer gegen alle Colizine der Gruppe wie auch gegen den Phagen BF 23 resistent. Demnach ist die Immunität durch Anwesenheit eines colizinogenen Faktors mehr spezifisch als die Resistenz durch Verlust eines Rezeptors.

2. Immunität und Resistenz verhalten sich wie nicht allelische Erbmerkmale. Die Kreuzung zweier Stämme, von denen der eine wegen der Anwesenheit eines colizinogenen Faktors und der andere wegen des Verlustes des Rezeptors gegen dasselbe Colizin unempfindlich sind, liefert Rekombinanten, die den Rezeptor von dem immunen Partner und die abwesende Immunität von dem resistenten Partner erworben haben und ganz empfindlich sind.

Diese Immunität ist je nach dem erzeugten Colizintyp mehr oder weniger ausgeprägt, ist aber nur selten total. Die für ein Colizin der Gruppe E transduzierten Stämme sind offensichtlich unempfindlich gegen das Colizin, das sie bilden. Jedoch verlieren sie oft im Laufe der Subkulturen den Rezeptor dieses Colizins, wahrscheinlich deshalb, weil spontane Mutanten, die diesen Rezeptor verlieren, schneller wachsen als die Zellen, die ihn behalten, und auf diese Weise selektioniert werden. Die für das Colizin I transduzierten Stämme werden nicht durch die Colizinmenge, die sie erzeugen, gehemmt, können aber gegen höhere Konzentrationen dieses Colizins empfindlich sein. Die für das Colizin V transduzierten Stämme haben schließlich eine noch schwächere Immunität und bleiben teilweise empfindlich gegen das Colizin selbst, das sie bilden. Ihre Kulturen in Bouillon sind weniger trübe als die des nichttransduzierten Zeugenstammes, die Stiche auf Agar entwickeln sich schlecht und die isolierten Kolonien sind klein und unregelmäßig [34]. Übrigens ist auch ein spontan colizinogener Stamm, der gegen das Colizin, das er bildet, empfindlich ist, nachgewiesen worden [84].

Die colizinogenen Faktoren sind also potentielle Letalfaktoren, deren Pathogenität nur nach der spontanen oder induzierten Entwicklung der Colizinsynthese, die sie steuern, in Erscheinung tritt.

V. Schlußfolgerungen

Mehrere *Coli-* oder verwandte Stämme können verschiedene, als Colizine bezeichnete antibiotische Stoffe bilden. Diese Colizine sind proteinische oder polypeptidische Stoffe, die durch zahlreiche Merkmale voneinander abweichen. Ihre Aktivitätsspektra sind sehr verschieden, aber streng auf andere Stämme derselben Familie beschränkt und dadurch bestimmt, daß die Colizine auf spezifische Rezeptoren der empfindlichen Bakterien zuerst haften müssen, um eine Hemmwirkung auszuüben.

Auf Grund ihrer Erbstabilität und ihrer Beständigkeit müssen die colizinogenen Eigenschaften durch genetische Faktoren bedingt sein, die die Synthese der Colizine steuern und ihre genetische Kontinuität sicherstellen. Diese genetischen colizinogenen Faktoren können in nichtcolizinogene Stämme durch Transduktion eingeführt werden. Obwohl die Übertragung der colizinogenen Eigenschaften den Kontakt und wahrscheinlich die Fusion zweier Zellen erfordert, ist sie doch von der sexuellen Rekombination der anderen Erbmerkmale ganz verschieden. Die colizinogenen Faktoren sind an keinen anderen genetischen Markierer gebunden und daher von der normalen genetischen Struktur der Bakterien unabhängig.

Colizin wird nicht von allen Zellen einer colizinogenen Kultur ausgeschieden, und die wenigen Zellen, wo die Colizinsynthese stattfindet, werden durch diese Synthese getötet. Diese tödliche Synthese kann in fast allen Zellen einer colizinogenen Kultur durch zahlreiche mutagene oder cancerigene Agenzien induziert werden. Die colizinogenen Faktoren sind also potentielle Letalfaktoren, deren Pathogenität nur nach der spontanen oder induzierten Entwicklung der Colizinsynthese enthüllt wird.

Auf Grund ihrer Übertragbarkeit, ihrer Unabhängigkeit von der normalen genetischen Struktur der Bakterien und ihrer Pathogenität könnten die colizinogenen Faktoren als Bakterienviren angesehen werden, und dies erklärt vielleicht die engen Beziehungen, die die Colizine mit Bakteriophagen aufweisen.

In gewisser Hinsicht verhalten sich wirklich die Colizine wie die Bakteriophagen und besitzen deren Spezifizität [10, 21, 25, 36, 70]. In beiden Fällen sind die Aktivitätsspektra durch spezifische Rezeptoren der empfindlichen Bakterien bedingt, und spezifische Colizine und Bakteriophagen haften auf ein und demselben Rezeptor [44]. Manche Bakteriophagen enthalten eine bakterizide Komponente, die den Colizinen verwandt ist [24, 73].

Während die Bakteriophagen biologische Einheiten sind, die durch die Zellen, die sie lysieren, reproduziert und vermehrt werden, sind jedoch die Colizine leblose chemische Stoffe, die die Zellen töten, ohne reproduziert zu werden. Die genetische Struktur, die in den Bakteriophagenpartikeln eingeschlossen ist und ihre Synthese steuert, fehlt den Colizinen, wird aber getrennt in den colizinogenen Bakterien fortgepflanzt. Der Übertragungsmechanismus der colizinogenen Faktoren muß also von dem der Bakteriophagen ganz verschieden sein. Die colizinogenen Faktoren werden nicht durch Lyse freigemacht und bleiben in den erzeugenden Bakterien, nur wird das Colizin in das äußere Milieu ausgeschieden. Aus Mangel an einem eigenen Übertragungsmechanismus ist das Überführen der colizinogenen Faktoren nur durch Fusion zweier Zellen möglich.

Antibiotika erzeugende virusähnliche Faktoren sind vielleicht auch in anderen Bakteriengruppen vorhanden. Gewisse *P. aeruginosa-* [61, 67] und

B. megaterium-Stämme [66] bilden spontan und in höherem Maß nach UV-Induktion sehr spezifische hochmolekulare antibiotische Stoffe, deren genetische Determinanten nicht näher bekannt sind. Auch bei tierischen Organismen, dem *Paramaecium*, sind genetische Faktoren, die in mancher Hinsicht wie Viren sich verhalten und die Bildung von spezifischen Hemmungsstoffen steuern, nachgewiesen worden [85]. Wie die colizinogenen Faktoren liegen diese anderen Faktoren in dem Grenzgebiet zwischen Plasmagenen und Viren, wo eine Demarkationslinie so schwer zu ziehen ist.

Literatur

[1] ABBOTT, J. D., and R. SHANNON, 1956: A method for typing *Shigella sonnei* using colicine production as a marker. J. Path. Bacter. **72**, 350.

[2] BLACKFORD, V. L., L. W. PARR, and M. L. ROBBINS, 1951: Antibiotic activity of selected enteric organisms. Antibiot. Chemother **1**, 392.

[3] BORDET, P., 1947: Propriétés distinctes d'antibiotiques produits par 2 souches de colibacille dont l'une dérive de l'autre. Rev. Immunol. **2**, 323.

[4] — 1948: Inhibition de l'action d'antibiotiques par les sérums antibactériens. C. r. Soc. Biol. **142**, 257.

[5] — et J. BEUMER, 1948: Inhibition de l'action d'antibiotiques par des extraits de bactéries sensibles. C. r. Soc. Biol. **142**, 259.

[6] — — 1951: Antibiotiques colibacillaires et récepteurs appropriés. Rev. belge Path. Méd. expér. **21**, 245.

[7] CHABBERT, Y., 1950: Analyse bactériologique des complexes de colicines produits par 14 souches d'*Escherichia coli* antibiotiques. Ann. Inst. Pasteur **79**, 51.

[8] COOK, M. K., V. L. BLACKFORD, M. L. ROBBINS, and L. W. PARR, 1953: An investigation of the antibacterial spectrum of colicines. Antibiot. Chemother. **3**, 195.

[9] DEPOUX, R., et Y. CHABBERT, 1953: Étude de l'activité antibiotique d'une colicine. Ann. Inst. Pasteur **84**, 798.

[10] Editorial, 1953: Colicines and bacteriophages. Lancet I, 1084.

[11] FASTIER, L. B., 1949: An antibiotic substance produced by a member of the *Shigella* group. J. Immunol. **62**, 399.

[12] FREDERICQ, P., 1946: Rapports entre les propriétés biochimiques et les phénomènes d'antibiose dans le groupe Coli-aerogenes. C. r. Soc. Biol. **140**, 1033.

[13] — 1946: Sur la spécificité des actions antibiotiques. Schweiz. Z. Path. Bakter. **9**, 385.

[14] — 1946: Sur la pluralité des récepteurs d'antibiose d'*E. coli*. C. r. Soc. Biol. **140**, 1189.

[15] — 1948: Production de substances antibiotiques par certaines souches de *Shigella*. C. r. Soc. Biol. **142**, 399.

[16] — 1948: Sur la destruction des colicines par les protéases microbiennes. C. r. Soc. Biol. **142**, 403.

[17] — 1948: Indépendance des mutations intéressant les récepteurs d'antibiose chez les Entérobactériacées. C. r. Soc. Biol. **142**, 855.

[18] — 1948: Actions antibiotiques réciproques chez les *Enterobacteriaceae*. Rev. belge Path. Méd. expér. **19**, suppl. **4**, 1.

[19] — 1950: Recherches sur les colicines du groupe E. C. r. Soc. Biol. **144**, 297.

[20] — 1950: Recherches sur les caractères et la distribution des souches productrices de colicine B. C. r. Soc. Biol. **144**, 1287.

[21] — 1950: Rapports entre colicines et bactériophages. Bull. Acad. roy. Méd. Belg. **15**, 491.

[22] — 1952: Recherche des propriétés lysogènes et antibiotiques chez les *Salmonella*. C. r. Soc. Biol. **146**, 298.

[23] — 1952: Action bactéricide de la colicine K. C. r. Soc. Biol. **146**, 1295.

[24] — 1952: Mise en évidence de 2 propriétés distinctes et indépendantes de la particule de phage T. 6. C. r. Soc. Biol. **146**, 1406.

[25] — 1953: Colicines et bactériophages. Ann. Inst. Pasteur **84**, 294.

[26] FREDERICQ, P., 1953: Recherches sur les caractères e la distributation des souches productrices de diverses colicines dans les selles normales et pathologiques. Bull. Acad. roy. Méd. Belg. **18**, 126.
[27] — 1953: Actions des colicines E et K sur la multiplication de divers bactériophages. C. r. Soc. Biol. **147**, 533.
[28] — Transduction génétique des propriétés colicinogènes chez *Escherichia coli* et *Shigella sonnei*. C. r. Soc. Biol. **148**, 399.
[29] — 1954: Rôle éventuel des propriétés lysogènes dans la transduction des propriétés colicinogènes chez *Escherichia coli*. C. r. Soc. Biol. **148**, 624.
[30] — 1954: Intervention du facteur de polarité sexuelle F dans la transduction des propriétés colicinogènes chez *Escherichia coli*. C. r. Soc. Biol. **148**, 746.
[31] — 1954: Induction de la production de colicine par irradiation ultra-violette de souches colicinogènes d'*Escherichia coli*. C. r. Soc. Biol. **148**, 1276.
[32] — 1955: Induction de la production de colicine et de bactériophages par irradiation ultraviolette de souches colicinogènes et lysogènes d'*Escherichia coli*. C. r. Soc. Biol. **149**, 2028.
[33] — 1956: Recherches sur la fréquence des souches transductrices des propriétés colicinogènes. C. r. Soc. Biol. **150**, 1036.
[34] — 1956: Résistance et immunité aux colicines. C. r. Soc. Biol. **150**, 1514.
[35] — 1957: Colicins. Ann. Rev. Microbiol. **11**, 7.
[36] — 1957: Colizine und Bakteriophagen. Protoplasma **48**, 4.
[37] — et M. BETZ-BAREAU, 1948: Caractères distinctifs des *E. freundii* isolées de l'urine des malades typhiques. C. r. Soc. Biol. **142**, 1180.
[38] — — 1952: Recombinants génétiques de souches marquées par résistance aux colicines et aux bactériophages. Ann. Inst. Pasteur **83**, 283.
[39] — — 1953: Transfert génétique de la propriété de produire un antibiotique. C. r. Soc. Biol. **147**, 1653.
[40] — — 1953: Transfert génétique de la propriété colicinogène en rapport avec la polarité F des parents. C. r. Soc. Biol. **147**, 2043.
[41] — — 1956: Influence de diverses propriétés colicinogènes sur la fertilité d'*Escherichia coli*. C. r. Soc. Biol. **150**, 615.
[42] — — et P. NICOLLE, 1956: Typage des souches d'*Escherichia coli* de gastroentérite infantile par recherche de leurs propriétés colicinogènes. C. r. Soc. Biol. **150**, 2039.
[43] — et G. DELCOUR, 1953: Sur la cinétique de l'action bactéricide des colicines E et K. C. r. Soc. Biol. **147**, 1310.
[44] — et A. GRATIA, 1949: Résistance croisée à certaines colicines et à certains bactériophages. C. r. Soc. Biol. **143**, 560.
[45] — E. JOIRIS, M. BETZ-BAREAU et A. GRATIA, 1949: Recherche de germes producteurs de colicines dans les selles de malades atteints de fièvre paratyphoïde B. C. r. Soc. Biol. **143**, 556.
[46] — and M. LEVINE, 1947: Antibiotic relationship among the enteric group of bacteria. J. Bacter. (Am.) **54**, 785.
[47] GARDNER, J. F., 1950: Some antibiotics formed by *Bacterium coli*. Brit. J. exper. Path. **31**, 102.
[48] GOEBEL, W. F., G. T. BARRY, M. A. JESAITIS, and E. M. MILLER, 1955: Colicine K. Nature **176**, 700.
[49] — — and T. SHEDLOVSKY, 1956: Colicine K. I. The production of colicine K in media maintained at constant pH. J. exper. Med. (Am.) **103**, 577.
[50] GRATIA, A., 1925: Sur un remarquable exemple d'antagonisme entre 2 souches de colibacille. C. r. Soc. Biol. **93**, 1040.
[51] — 1932: Antagonisme bactérien et bactériophagie. Ann. Inst. Pasteur **48**, 113.
[52] — P. FREDERICQ, E. JOIRIS, M. BETZ-BAREAU et E. WEERTS, 1950: La technique de la double couche de gélose et la distribution de germes producteurs de colicines dans les matières fécales normales et pathologiques de l'homme et des animaux. Ant. van Leeuwenhoek **16**, 31.
[53] GROSSO, E., 1950: Contributo allo studio delle colicine. Boll. Ist. sieroter. milan. **29**, 373.
[54] HALBERT, S. P., 1948: The antagonism of coliform bacteria against *Shigella*. J. Immunol. **58**, 153.
[55] — 1948: The relation of antagonistic coliform organisms to *Shigella* infections. I. Survey observations. J. Immunol. **60**, 23.
[56] — 1948: The relation of antagonistic coliform organisms to *Shigella* infections. II. Observations in acute infections. J. Immunol. **60**, 359.

[57] Halbert, S. P., and M. Gravatt, 1949: The relation of *Shigella* type specificity and susceptibility to antibiotic producing strains of *E. coli.* J. Immunol. **61**, 271.
[58] — — 1949: Prevalence of antibiotic producing coliform organisms. Publ. Health Rep. (Am.) **64**, 313.
[59] — and H. J. Magnuson, 1948: Studies with antibiotic producing strains of *E. coli.* J. Immunol. **58**, 397.
[60] Hamon, Y., 1955: Étude d'une colicine élaborée par une culture de *S. paratyphi* B. Ann. Inst. Pasteur **88**, 193.
[61] — 1956: Contribution à l'étude des pyocines. Ann. Inst. Pasteur **91**, 82.
[62] — 1956: Les bactériocines sont-elles des substances antigéniques? C. r. Acad. Sci. **242**, 1240.
[63] — 1957: Propriétés générales des cultures d'entérobactériacées rendues colicinogènes. Ann. Inst. Pasteur **92**, 363.
[64] — et Z. Lewe, 1955: Étude de l'induction par irradiation ultraviolette de quelques cultures d'*E. coli* K. 12 préalablement rendues colicinogènes par transduction. Ann. Inst. Pasteur **89**, 336.
[65] Heatley, N. G., and H. W. Florey, 1946: An antibiotic from *Bacterium coli.* Brit. J. exper. Path. **27**, 378.
[66] Ivanovics, G., and L. Alföldi, 1954: A new antibacterial principle: Megacine. Nature **174**, 465.
[67] Jacob, F., 1954: Biosynthèse induite et mode d'action d'une pyocine, antibiotique de *Pseudomonas pyocyanea.* Ann. Inst. Pasteur **86**, 149.
[68] — L. Siminovitch, et E, Wollman, 1951: Induction de la production d'une colicine par le rayonnement ultraviolet. C. r. Acad. Sci. 233, 1500.
[69] — — — 1952: Sur la biosynthèse d'une colicine et sur son mode d'action. Ann. Inst. Pasteur **83**, 295.
[70] — — — 1953: Comparaison entre la biosynthèse induite de la colicine et des bactériophages et entre leur mode d'action. Ann. Inst. Pasteur **84**, 313.
[71] Jenkin, C. R., and D. Rowley, 1955: Resistance to colicine E as a genetic marker in coli K. 12. Nature **175**, 779.
[72] Kellenberger, G., et E. Kellenberger, 1956: Étude de souches colicinogènes au microscope électronique. Schweiz. Z. Path. Bakter. **19**, 582.
[73] Latarjet, R., and P. Fredericq, 1956: An X-ray study of a colicine and of its relationship to bacteriophage T. 6. Virology **1**, 100.
[74] Lederberg, J., 1947: Gene recombination and linked segregations in *E. coli.* Genetics **32**, 505.
[75] — 1952: Cell genetics and hereditary symbiosis. Physiol. Rev. **32**, 403.
[76] — 1955: Recombination mechanisms in bacteria. J. cellul. a. comp. Physiol. (Am.) **45**, 75.
[77] Levine, M., and R. H. Tanimoto, 1954: Antagonisms among enteric pathogens and coliform bacteria. J. Bacter. (Am.) **67**, 537.
[78] Ludford, C. G., and M. Lederer 1953: The antibiotics of *Escherichia coli.* Austral. J. exper. Biol. a. med. Sci. **31**, 553.
[79] Lwoff, A., et F. Jacob, 1952: Induction de la production de bactériophages et d'une colicine par les peroxydes, les ethyleneimines et les halogénoalcoylamines. C. r. Acad. Sci. **234**, 2308.
[80] Mondolfo, U., 1948: Sulla produzione di sostanze antibiotiche da parte dell'*E. coli.* Boll. Soc. ital. Biol. sperim. **24**, 1.
[81] Nozaki, J. N., M. L. Robbins, and L. W. Parr, 1953: The fundamental nature of colicine production. J. Bacter. (Am.) **66**, 621.
[82] Panijel, J., et J. Huppert, 1956: Colicines et prolysines. C. r. Acad. Sci. **242**, 199.
[83] Robbins, M. L., N. N. El Shawi, and L. W. Parr, 1956: Colicine susceptibility patterns of enteropathogenic strains of *Escherichia coli.* Bacter. Proc. **97**.
[84] Ryan, F. J., P. Fried, and F. Mukai, 1955: A colicin produced by cells that are sensitive to it. Biochim. Biophys. Acta **18**, 131.
[85] Sonneborn, T. M., 1946: Experimental control of cytoplasmic factors in *Paramecium.* Cold Spring Harbor Symp. Quant. Biol. **11**, 236.

GPSR Compliance

The European Union's (EU) General Product Safety Regulation (GPSR) is a set of rules that requires consumer products to be safe and our obligations to ensure this.

If you have any concerns about our products, you can contact us on ProductSafety@springernature.com

In case Publisher is established outside the EU, the EU authorized representative is:

Springer Nature Customer Service Center GmbH
Europaplatz 3
69115 Heidelberg, Germany

Batch number: 10578397

Printed by Printforce, the Netherlands